Jay Kalra

Medical Errors and Patient Safety

# Patient Safety

Edited by
Oswald Sonntag and Mario Plebani

**Volume 1**

Jay Kalra

# Medical Errors and Patient Safety

Strategies to reduce
and disclose medical
errors and improve
patient safety

**DE GRUYTER**

**Author**

Jay Kalra, MD, PhD, FRCPC, FCAHS
Department of Pathology and Lab Medicine
College of Medicine
University of Saskatchewan
Royal University Hospital
103 Hospital Drive
Saskatoon
Saskatchewan S7N 0W8
Canada

ISBN 978-3-11-024949-1 • e-ISBN 978-3-11-024950-7

*Library of Congress Cataloging-in-Publication Data*

Kalra, Jay.
    Medical errors and patient safety / by Jay Kalra.
        p. ; cm. — (Patient safety)
    ISBN 978-3-11-024949-1 (alk. paper)
    1.  Medical errors—United States.   2.  Patient safety.   3.  Patients—Safety measures—
United States.   I. Title.   II. Series: Patient safety.
    [DNLM: 1.  Medical Errors. 2.   Safety Management.   WB 100]

    R729.8.K35 2011
    610.28'9—dc22                                                          2011002654

*Bibliographic information published by the Deutsche Nationalbibliothek*
The Deutsche Nationalbibliothek lists this publication in the Deutsche Nationalbibliografie; detailed
bibliographic data are available in the Internet at http://dnb.d-nb.de.

Project management: Dr. Petra Kowalski.
Production editor: Heike Jahnke.
Production manager: Ulrike Swientek.
Typesetting: Apex CoVantage, LLC
Cover image: Comstock/Getty Images.
Printing and binding: Hubert & Co GmbH & Co KG, Göttingen.

Printed in Germany
⊗ Printed on acid-free paper.

www.degruyter.com

**Dedicated to**
My wife and children, for teaching me the value of family

# Contents

# Acknowledgments

This book comes out of ongoing work at the University of Saskatchewan and Royal University Hospital, Saskatoon Health Region, Saskatoon, Saskatchewan, and related projects devoted to addressing medical errors, patient safety, and quality programs across the fields of medicine and health care.

I wish to express my sincere gratitude to my friends and colleagues at the Department of Pathology and Laboratory Medicine at the University of Saskatchewan and Saskatoon Health Region who have offered support and valuable assistance to this project. This project has benefited from helpful discussions at various scientific meetings, panel presentations, and colloquiums and from the thoughts of students in different classes. Gratitude must be extended to Dr. H. E. Emson, Professor (Emeritus) and former Department Head, for his constant guidance, encouragement, and personal concern during the early stages of my career. I wish to thank Diane Collard, Annamae Giles, Cheryl Booth, Lyndon Entwistle, Maxim Gertle-Jaffe, Ajay Nayar, and Amith Mulla for their helpful suggestions, as well as for their interest in this work. The author would like to take this opportunity to thank Heather Neufeld, and Lisa Worobec for their professional work diligence in reviewing the complete manuscript at different stages and in offering critical suggestions and comments. My thanks are also due to Bill Gray and Todd Reichert for their assistance in the art work and graphic design.

# About the author

Dr. Jay Kalra, an educator, researcher and quality health care advocate, is a Professor of Pathology at the University of Saskatchewan and has served as Head of the Department of Pathology and Head of the Department of Laboratory Medicine, Saskatoon District Health. He is a Fellow of the Royal College of Physicians and Surgeons of Canada (FRCPC), the Canadian Academy of Clinical Biochemistry (FCACB), the Canadian Academy of Health Science (FCAHS), and an Elected Fellow of the Royal Society of Medicine, UK. Dr. Kalra has served as the President of the Canadian Association of Medical Biochemists, the Intersociety Council of Laboratory Medicine of Canada, the Canadian Chairs of Pathology and Laboratory Medicine, the Canadian Association of Pathologists and as the Director of Saskatchewan Stroke Research Centre.

Jay is a pioneer in establishing guidelines for thyroid-function testing, quality assurance programs and laboratory utilization in health care. Dr Kalra is advancing the agenda, nationally and internationally, related to quality care and patient safety including risk management, ethical issues, disclosure of medical/clinical error issues and related policies and practices.

# Abbreviations

| | |
|---|---|
| ACGME | Accreditation Council for Graduate Medical Education |
| AIMS | Australian Incident Monitoring System |
| AMA | American Medical Association |
| AMI | Acute Myocardial Infarction |
| CAP | College of American Pathologists |
| CLIA | Clinical Laboratory Improvement Act |
| CQI | Continuous Quality Improvement |
| CPSI | Canadian Patient Safely Institute |
| DMAIC | Define, Measure, Analyze, Improve, Control |
| DMADV | Define, Measure, Analyze, Design, Verify |
| DPM | Defects per Million |
| ECG | Electrocardiograph |
| ED | Emergency Department |
| EQA | External Quality Assessment |
| GE | General Electric |
| HMP | Harvard Medical Practice |
| ICU | Intensive Care Unit |
| IOM | Institute of Medicine |
| JCAHO | Joint Commissions on Accreditation of Health Care Organization |
| NHS | National Health Service |
| NICU | Neonatal Intensive Care Unit |
| NPSD | Network of Patient Safety Database |
| PICU | Pediatric Intensive Care Unit |
| POCT | Point-of-Care Testing |
| PSO | Patient Safety Organization |
| PT | Proficiency Testing |
| QAHC | Quality of Australian Health Care |
| QAHCS | Quality of Australian Health Care Study |
| QC | Quality Control |
| SD | Standard Deviation |
| SHN | *Safer Healthcare Now!* |
| SICU | Surgical Intensive Care Unit |
| TQM | Total Quality Management |
| TAT | Turnaround Times |
| TSH | Thyroid Stimulating Hormones |

# 1 An overview and introduction to concepts

The prevalence of medical errors in health care systems has generated immense interest in recent years, despite being ever-present in health care delivery to some degree. The research has consistently revealed high rates of adverse events in hospitalized populations. Some of these adverse events result from medical errors, and a majority of these errors may be preventable. These errors can occur anywhere and at any time in health care processes. The consequences of these errors may vary from little or no harm to patient fatalities. It is important to recognize that a degree of error is inevitable in any human task, and human fallibility in health care should be accepted. The underlying precursors for many of these human errors may be latent systemic factors inherent in today's increasingly complex health care system. The focus of adverse event analyses on individual shortcomings without appropriate attention to system issues leads to ineffective solutions. The cognitive influence on medical decision making and error generation is also significant and should not be discounted.

## 1.1 Introduction

Health care processes are increasingly being implicated in causing harm to patients. Medical errors and adverse events are primarily responsible for this harm. These errors, which may occur at every level of the system, are both common and diverse in nature. The magnitude and potential for errors is enormous and is increasingly threatening to become a regular feature of the health care system.

It was precisely with these concerns in mind that Hippocrates coined the maxim "First, do no harm." Though it was phrased more than 2,000 years ago, can anyone question its wisdom and relevance today? A little more than a decade ago, it was generally perceived that adverse outcomes from medical care or interventions were relatively rare and isolated incidents. Patients enjoyed a perception that health care professionals performed nearly perfectly in a perfect health care system.

In recent times, however, this sense of perceived safety within the confines of health care processes has been challenged. In a survey commissioned by the National Patient Safety Foundation at the American Medical Association (AMA), the general public perceived the health care environment as "moderately" safe. The survey further indicated that the health care environment was rated a 4.9 on a scale of 1 to 7, where 1 is not safe at all and 7 is very safe. The health care environment fared badly in comparison with airline travel and workplace safety, which were rated higher. There are other reports in the medical literature that indicate that every patient subjected to medical intervention may not necessarily face a beneficial outcome (Brennan et al., 1991; Kohn et al., 2000; Vincent et al., 2001; Wilson et al., 1995). This finding is not new and was reported close to four decades ago by Schimmel (1964). In 2000, the Institute of Medicine (IOM) report *To Err Is Human: Building a Safer Health System* went to the extent of terming medical errors a public health risk (Kohn et al., 2000).

## 1.2 Medical error

The term "error" has been variously defined. The Oxford Dictionary of Current English (1998) defines it as "mistake" or the condition of being morally "wrong"; a little more philosophically William Ellery Channing, a 19th-century essayist, noted that "Error is discipline through which we advance." Error has also been defined in a wider context as being a problem in the process of care itself (Hofer et al., 2000) or as the failure of a planned action to be completed as intended or the use of a wrong plan to achieve an aim (Reason, 1990). Although the definition of "error" has its origins in behavioral psychology, the term is appropriate for medical usage. Using Reason's definition, IOM has tried to separate medical error into two parts (Kohn et al., 2000): the first half of the definition constitutes "error of execution" and the latter half, "error of planning." In this context, two other related terms, "adverse event" and "patient safety," should be defined and elaborated. Bates et al. (1997) defined adverse events as injuries that result from medical management, rather than from the underlying disease. Patient safety, as defined by IOM, is freedom from accidental injury (Kohn et al., 2000).

All three terms, "medical error," "adverse event," and "patient safety," complement one another. However, special emphasis should be placed on the terms "medical error" and "adverse event." Adverse events have been widely classified as preventable and unpreventable, though some suggest that a preventable adverse event constitutes an error (Leape, 1994; Leape et al., 1995) and a subset of adverse events can be judged preventable, though preventability has never been rigorously measured (Hofer et al., 2000). There are many questions that continue to persist concerning the actual relationship between errors and adverse events (Brennan, 2000; Hofer et al., 2000). More important, it must be mentioned that errors should not be considered equivalent to negligence (Vincent, 1989). The former has a large circumstantial component, whereas the latter is a reflection of true incompetence in medical practice. The tort law defines medical negligence as a failure to meet the standard of practice of an average qualified physician practicing the specialty in question (Brune v. Belinkoff, 1968). Though constituting a significant part of preventable errors and adverse events, negligence is generally considered an extreme degree of error.

In some cases, adverse events occur in the absence of errors or negligence, but patient safety is nonetheless compromised. In such cases, it is the inevitability of an event that is held to be accountable. In lieu of these limitations, it is suggested that future research should use observational and epidemiological studies incorporating various methods to ascertain, compare, and validate all components of medical error, including the degree of harm, preventable adverse events, near misses, errors in the process, and the error-free functioning of the health care system (Elder and Dovey, 2002).

Another difficulty in differentiating medical errors from adverse events arises in the context of the eventual outcome of a medical action. It is worth asking why a perfect process carried out with all due care and proper precautions may end with an adverse outcome or, alternately, why sometimes errors are recognized only retrospectively after an adverse outcome. There exists a simple explanation: a perfectly reasonable action may be labeled erroneous if it results in an adverse outcome. This dilemma may not have simple answers, nor is answering this question the principal objective; the objective instead is reducing the incidence of adverse outcomes by practicing more patient-safe medicine.

Because of the complexity of health care and the current rate of adverse events, it must be mentioned that health care cannot be compared with other high-risk industries (see ▶Fig. 1.1), such as aviation and nuclear energy, both of which boast lower levels of risk (Amalberti et al., 2005; Law, 2004; Leape, 1994). However, because of the safer outcomes of these other sectors, they may be useful industries to borrow from in terms of safety programs and policies. Reason (1990) noted that health care is being delivered in an environment in which there are complex interactions among many variables, such as the disease process itself, the medical staff and equipment, the infrastructure, and organizational policies and procedures. Continuous advances in the field require that medical professionals constantly familiarize themselves with new equipment, pharmaceuticals, and methods. Moreover, each component of the medical process is intrinsically complicated at the same time that the performance of each part is affected by the functioning of other component parts within the system. Thus, even small errors can result in larger system consequences. Unlike many other industries, health care sectors do not enjoy the luxury of well-defined processes. Health care professionals function in a dynamic environment. Some health care services, such as the emergency department (ED), where aggressive interventions are often necessary, suffer the most from these liabilities. Some of the key decisions made in EDs are taken in split seconds and are frequently based on little or no prior medical information. It is therefore not surprising that the rates of error are the highest in such challenging environments.

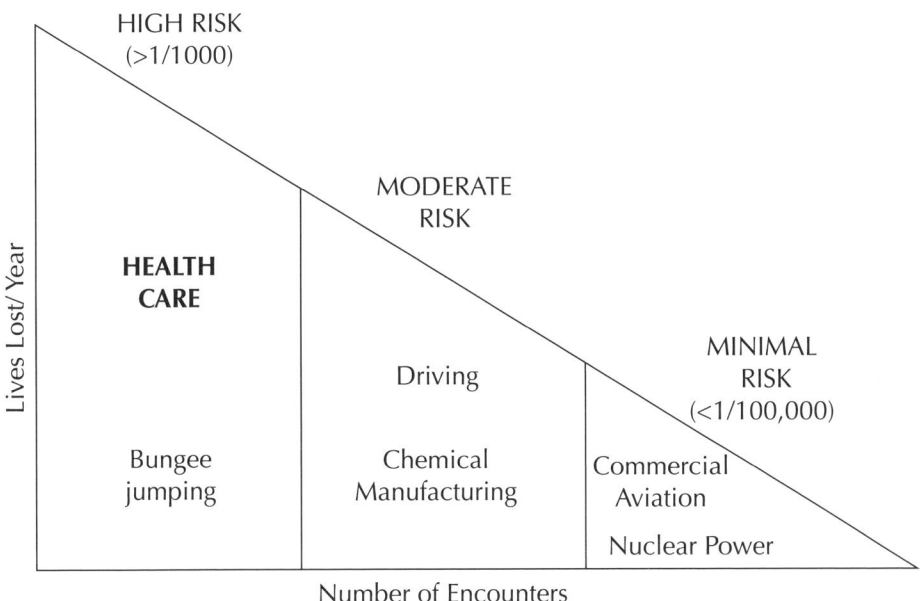

**Fig. 1.1:** Comparison of risk in health care and other industries.

The questions remaining to be answered are, in essence, these: what are the causes for these errors, and what can be done to prevent them? The promotion of patient safety may require greater integration of concepts evolving from cognitive psychology, human and systems re-engineering, bioethical viewpoints, social perception, and legislative actions. This book intends to address such topics.

## 1.3 Magnitude and epidemiology of health care errors

Research on medical errors has consistently revealed high error rates across the world (see ▶Tab. 1.1). The IOM report estimated that between 44,000 and 98,000 deaths occur every year in the United States as a result of medical error (Kohn et al., 2000). This report was extensively covered by the media and raised public awareness of issues related to the quality of health care and the safety of patients in the United States. The report prompted immediate presidential and congressional action (Charatan, 2000). Schimmel (1964), in a university hospital–based study, observed that at least 20% of hospitalized patients suffered an iatrogenic injury, which is an adverse event caused by medical examination or treatment. This study by Schimmel in 1964, one of the first published in the field, was carried out in a medical era different from the present one and may not be an accurate reflection of the improved standards in care that are seen today. It nevertheless highlights the concerns related to adverse events and patient safety of that time. Steel et al. (1981) estimated a much higher rate of iatrogenic injuries, asserting that 36% of admitted patients suffered from such an injury. However, the high adverse event rate reported by Steel et al. included incidents of patient falls, which may be difficult to control. A decade later, Brennan et al. (1991) published the findings of their landmark Harvard Medical Practice (HMP) study. This was a population-based study of medical injuries caused by interventions. In this study, the authors reviewed medical charts of more than 30,000 patients admitted to New York hospitals in 1984 and estimated that 3.7% of admissions resulted in an injury that prolonged the patient's hospital stay or resulted in disability at the time of discharge. Extrapolation of the HMP study data translated into an estimated nearly 100,000 patients suffering from an iatrogenic injury in 1984 in New York alone. Though the HMP study estimated that iatrogenic injuries occurred at rates much lower than those found in the two previous studies, the magnitude of the numbers quoted by the study was high, forcing many to ponder the standards of patient safety in present-day health care. The Californian Medical Insurance Feasibility Study of 1977 found that 4.6% of 21,000 hospitalizations reviewed resulted in iatrogenic injury (California Medical Association, 1977).

The concern with health care quality and patient safety is not unique to the United States. The Quality of Australian Health Care (QAHC) Study, another population-based study that soon followed, estimated that 16.6% of admissions resulted in adverse events (Wilson et al., 1995). In terms of seriousness of injury, the HMP and QAHC studies differed widely. The HMP study reported that approximately 14% of the injuries were ultimately fatal and approximately 3% caused permanent disability, whereas the QAHC study estimated that death occurred in nearly 5% of injuries and permanent disability in 14% of patients who suffered from iatrogenic injuries. The HMP study was followed up with a similar study from the states of Utah and Colorado (Thomas et al., 2000). The Utah and Colorado study employed a representative sample of around 15,000

**Tab. 1.1:** Major studies on rates of adverse outcomes in health care.

| Study | Year | Data Source | Rate of Adverse Outcomes | Fatality Rates | Prime Events of Concern |
|---|---|---|---|---|---|
| Schimmel | 1964 | 1,252 admissions in a university hospital | 20% | 8% | Medication and diagnostic procedures |
| Steel et al. | 1981 | 815 admissions in a university hospital | 36% | 2% | Medication and diagnostic procedures |
| Brennan et al. | 1991 | 30,121 records from New York hospitals | 3.7% | 13.6% | Medication and surgical procedures |
| Wilson et al. | 1995 | 14,000 admissions in New South Wales and South Australia | 16.6% | 4.9% | Medication and surgical procedures |
| Thomas et al. | 2000 | 15,000 non-psychiatric discharges in Utah and Colorado states | $2.9 \pm 0.2\%$ Mean $\pm 2SD$ | $6.6 \pm 1.2\%$ Mean $\pm 2SD$ | Medication and surgical procedures |
| Vincent et al. | 2001 | 1,014 records from two hospitals in London area | 10.8% | 8% | Surgical procedures |
| Forster et al. | 2004 | 620 patients from Canadian teaching hospital | 12.7% | 4.8% | Medication and operative complication |
| Baker et al. | 2004 | 25 hospitals, 3745 hospital charts | 7.5% | 20.8% | Diagnostic and surgical procedures |

nonpsychiatric discharges and found adverse events in about 3% of hospitalizations in each state.

These major studies were followed by a preliminary estimate of adverse events occurring in British hospitals (Vincent et al., 2001). Vincent et al. (2001), using a similar methodology of retrospective record review, reported that approximately 11% of patients in their study experienced an adverse event. The proportion of patients with permanent impairment or in whom the adverse event contributed to death was 6% and

8%, respectively. In New Zealand hospitals, an adverse event was recorded in at least 10.7% of the 1,326 medical records reviewed (Davis et al., 2001). Although the results were generated by an audit study of three public hospitals for admissions in 1995, these results were comparable with the QAHC study. In a study by O'Hara and Carson (1997) done in Australia on routine patient data collection, adverse events were reported in 5% of separations, with the incidence increasing with patient age. Schioler et al. (2001) studied the incidence of adverse events in Denmark and reported that 9% of all admissions were associated with adverse events. Though the adverse events recorded in the study caused a prolonged hospital stay, most resulted in minor and transient disabilities, while a few deaths or permanent disability were recorded. Another study based on data from New Zealand hospitals reported that the proportion of hospital admissions associated with an adverse event was nearly 13% (Davis et al., 2002).

Another study evaluated the effects of adverse events on patients after discharge from the general medical service of a tertiary-care academic hospital (Forster et al., 2003). Forster et al. (2003), in a prospective cohort study, interviewed a total of 400 patients an average 24 days postdischarge. The authors reported that at least 76 patients (19%) had symptoms from medical care. They further classified adverse events into two types: preventable events caused by an error and ameliorable events, in which the severity of the event could have been decreased. Each of these subtypes accounted for 6% of the total incidence reported. The authors emphasized the importance of focusing on ameliorable events, as these events, though unpreventable, had the potential of being less severe provided early corrective actions were initiated. It was also suggested that these ameliorable events were relevant to patient safety, especially in the postdischarge period.

The adverse events recorded in the previously discussed studies are widely distributed and are not restricted to any particular specialty or subspecialty. A careful review of the literature, however, indicates that procedure-related and drug-related events were the most risky for patients. The Harvard Medical Practice Study II reviewed the nature of adverse events in hospitalized patients in New York State in 1984 (Leape et al., 1991). The authors of this study reported that nearly half of all adverse events were associated with an operation. The authors also reported that drug-related complications were the single most common type of adverse event. The QAHC study reported that more than half of all the adverse events recorded were associated with an operation and nearly 11% of adverse events were attributed to drug-related complications. Other studies have reported similar findings (Davis et al., 2001; Thomas et al., 2000; Vincent et al., 2001). Gawande et al. (1999), however, observed that adverse events were as common in nonsurgical care as in surgical care. Though 66% of all adverse events in the study by Gawande et al. were related to surgical care, it was suggested that this might be the result of a high number of surgical encounters or consultations in admitted patients. Another study done with the sole objective of studying adverse event rates for surgical patients in Australia found a rate of approximately 22% for surgical admissions, with 13% of these patients suffering permanent disability and 4% of the events having fatal consequences (Kable et al., 2002).

In Canada, studies conducted by Baker et al. (2004) and Forster et al. (2004) detected that 7.5% and 12.7% of hospitalized patients were affected by an adverse event, respectively, and they deemed 2.8% and 4.8%, respectively, of these adverse events to have been preventable. In Baker et al.'s study, the most common causes of medical error were delayed diagnoses of cancer and heart disease, drug overdoses, communication errors, and operative errors. According to Forster et al. (2004), the principal causes were erroneous delivery of medication, operative complications, and nosocomial infections.

There are two other areas on which the epidemiological research on adverse events and medical error agree: that these events are preventable and that the frequency of adverse events rises with advanced patient age. With respect to preventability of the adverse events, a majority of the authors suggested that the adverse events were indeed preventable (Gawande et al., 1999; Leape et al., 1991; Vincent et al., 2001; Wilson et al., 1995). Davis et al. (2001), however, had a different finding. They reported that for 60% of adverse events the evidence for preventability was either low or nonexistent. The other wide point of consensus in the literature on adverse events and patient safety is that advanced age serves as a risk factor (Brennan et al., 1991; Kable et al., 2002; O'Hara and Carson, 1997; Wilson et al., 1995). Only one study failed to demonstrate this risk (Davis et al., 2002). Thomas and Brennan (2000) exclusively studied the incidence and types of preventable adverse events in elderly patients. These investigators estimated that adverse events occurred in 5.29% of all of elderly patients discharged in the two states of Utah and Colorado but in only 2.8% of nonelderly patients. Moreover, the results of the study revealed that preventable adverse events were more common in the elderly than in their younger counterparts. Many factors contribute to this increased incidence of adverse events among the elderly – chiefly their comorbid illnesses, complexity of care, and increased exposure to risk.

Although the reported studies may have applied rigorous methodologies and precisely defined adverse events, various shortcomings in their study methods can be identified. One of the larger studies (Thomas et al., 2000) and some smaller studies (Davis et al., 2002; O'Hara and Carson, 1997) that estimated rates of adverse events remained silent on the important factor of preventability. It would have been appropriate for the studies to include well-defined criteria for which adverse events were considered "highly preventable," "moderately preventable," and "totally unpreventable" in their report of their findings. These categories should have been assessed for inter-reviewer reliability to minimize the subjective bias inherent in many of these studies. It is important to note that only some studies included assessment of preventability of adverse events as part of the original study design (Davis et al., 2001; Schioler et al., 2001; Vincent et al., 2001; Wilson et al., 1995). In at least two of these studies, more than half of the adverse events were judged to be preventable (Vincent et al., 2001; Wilson et al., 1995). One of the studies, though, as acknowledged by the authors themselves, was a small pilot study made up of cases mainly from high-risk surgical specialties, which may have higher rates of adverse events than other specialties (Vincent et al., 2001). The HMP Study II data were analyzed for preventability of adverse events, and the authors suggested that many of the adverse events were neither preventable nor predictable based on the current state of medical knowledge (Leape et al., 1991).

Sox and Woloshin (2000) also expressed doubts about the methods used by Thomas et al. (1999) in classifying nearly half the adverse events as preventable as their study reviewed the summary of the events and not the entire medical record. The methods used to determine the effect of adverse events on eventual patient outcomes are also questionable. The IOM report based its findings on two studies, the HMP study and the study by Thomas et al. (2000). The data obtained in both studies were derived from hospitalized patients, who were at greater risk of health problems or death than the nonhospitalized population. Also, some studies (Brennan et al., 1991; Wilson et al., 1995) have reported large fatality rates due to adverse events but failed to explain whether the patients would have lived longer in the absence of adverse events.

Hayward and Hofer (2001), whose study reported a 6% fatality rate due to suboptimal care, recently evaluated this aspect. They noted, however, that, after consideration of the prognosis and other factors in these patients, they estimated that only 0.5% of the deceased would have gone on to live for three months or more. These findings can be questionable on an ethical basis. The high figures cited in the IOM report have also been previously questioned and termed unsubstantiated (Sox and Woloshin, 2000). These studies failed to achieve conclusiveness in demonstrating that the adverse events resulted directly from the error. The generalization of the rate of adverse events across North America may also be impractical, particularly in view of recent findings that sociodemographic variables may modify the risk for complications of medical care (Villanueva and Anderson, 2001). The other limitation in many of these studies is that their data were primarily based on hospitalized care and adverse events that were documented only in medical records. Many other patient care areas, including physician offices and outpatient departments, have not been considered.

It is unfair to suggest solely on the basis of rates of adverse events that patient safety has deteriorated over the past few decades. If the rate of adverse events is an indicator of patient safety, then the indications are that patient safety is steadily attaining greater levels. This can be assumed because the HMP Study I, which reviewed more than 30,000 hospital records from 1984, reported an adverse event incidence of 3.7%, whereas older studies gave much higher rates, varying from 20% in the 1960s (Schimmel, 1964) to 4.6% of all hospitalizations in the 1970s (California Medical Association, 1977). There may be some limitations to the HMP Study I, however, as it accounted only for adverse events that occurred in hospitalized care and those documented in medical records. Some have questioned the actual number of deaths due to adverse events and substituted their own estimates (Hayward and Hofer, 2001). Such statistics demonstrate a need for improvement – despite their uneven conclusions – and reflect an increase in interest and attention to patient safety.

## 1.4 Conclusion

Setting statistics and figures aside, the core of the issue remains that patient safety is at peril in today's health care system. Protecting patients from iatrogenic harm is not only the principal duty of every health care professional but also an ethical and moral responsibility. The responsibility is owed simply because of the trust placed by patients in health care professionals. This is a period of crisis and horrendous problems for health care. The debate is neither about the number of fatalities or injuries nor about the cost estimates of adverse events. It should instead be focused on how best to usher in a culture of safety by changing mindsets, understanding error mechanisms, and devoting resources towards improvements in this area.

Present and past approaches to patient welfare have demonstrated serious oversight in, and a blindness of faith toward, medical processes and professionals. Across contexts – nations and communities – it is apparent that individuals have only limited affect upon systemic issues. These are what need to be addressed.

The reviewed literature on adverse events may not truly reflect the current state of health care quality and patient safety issues because of some of the methodological flaws outlined earlier. However, the core issue remains that current health care processes are not performing at their potential best. There is tremendous scope for quality improvement in health care delivery.

## References

Amalberti R, Auroy Y, Berwick D, Barach P. Five system barriers to achieving ultrasafe health care. Ann Intern Med 2005;142:756–764.

Baker GR, Norton PG, Flintoft V, et al. The Canadian Adverse Events Study: the incidence of adverse events among hospital patients in Canada. CMAJ 2004;170:1678–1686.

Bates DW, Spell N, Cullen DJ, et al. The costs of adverse drug events in hospitalized patients. Adverse drug events prevention study group. JAMA 1997;277:307–311.

Brennan TA. The institute of medicine report on medical errors – could it do harm? N Engl J Med 2000;342:1123–1125.

Brennan TA, Leape LL, Laird NM, et al. Incidence of adverse events and negligence in hospitalized patients. Results of the Harvard Medical Practice Study I. N Engl J Med 1991; 324:370–376.

Brune v. Belinkoff, 354 Mass. 102 (1968).

California Medical Association. Report of the Medical Insurance Feasibility Study. San Francisco: The Association; 1977.

Charatan F. Clinton acts to reduce medical mistakes. BMJ 2000;320:597.

Davis P, Lay-Yee R, Briant R, Ali W, Scott A, Schug S. Adverse events in New Zealand public hospitals I: occurrence and impact. N Z Med J 2002;115:U271.

Davis P, Lay-Yee R, Schug S, et al. Adverse events regional feasibility study: indicative findings. N Z Med J 2001;114:203–205.

Elder NC, Dovey SM. Classification of medical errors and preventable adverse events in primary care: a synthesis of the literature. J Fam Pract 2002;51:927–932.

Forster AJ, Clark HD, Menard A, et al. Adverse events among medical patients after discharge from hospital. CMAJ 2004;170:345–349.

Forster AJ, Murff HJ, Peterson JF, Gandhi TK, Bates DW. The incidence and severity of adverse events affecting patients after discharge from the hospital. Ann Intern Med 2003;138:161–167.

Gawande AA, Thomas EJ, Zinner MJ, Brennan TA. The incidence and nature of surgical adverse events in Colorado and Utah in 1992. Surgery 1999;126:66–75.

Hayward RA, Hofer TP. Estimating hospital deaths due to medical errors: preventability is in the eye of the reviewer. JAMA 2001;286:415–420.

Hofer TP, Kerr EA, Hayward RA. What is an error? Eff Clin Pract 2000;3:261–269.

Kable AK, Gibberd RW, Spigelman AD. Adverse events in surgical patients in Australia. Int J Qual Health Care 2002;14:269–276.

Kohn LT, Corrigan JM, Donaldson MS, editors. Committee on Quality of Healthcare in America, Institute of Medicine. To Err Is Human: Building a Safer Health System. Washington, DC: National Academy Press; 2000.

Law R. Safety of dietary supplements: Figure 5: How safe is safe enough? http://www.laleva.cc/petizione/english/ronlaw_eng.html. Accessed November 1, 2010.

Leape LL, Error in medicine. JAMA 1994;272:1851–1857.

Leape LL, Bates DW, Cullen DJ, et al. Systems analysis of adverse drug events. ADE prevention study group. JAMA 1995;274:35–43.

Leape LL, Brennan TA, Laird N, et al. The nature of adverse events in hospitalized patients. Results of the Harvard Medical Practice Study II. N Engl J Med 1991;324:377–384.

O'Hara DA, Carson NJ. Reporting of adverse events in hospitals in Victoria, 1994–1995. Med J Aust 1997;166:460–463.

Oxford Dictionary of Current English. Thompson D, editor. New revised ed. Oxford: Oxford Univ. Press; 1998:294.

Reason JT. Human Error. Cambridge: Cambridge Univ. Press; 1990.

Schimmel EA. The hazards of hospitalization. Ann Intern Med 1964;60:100–110.

Schioler T, Lipczak H, Pedersen BL, et al. Incidence of adverse events in hospitals. A retrospective study of medical records. Ugeskr Laeger 2001;163:5370–5378 [Article in Danish].

Sox Jr. HC, Woloshin S, How many deaths are due to medical error? Getting the number right. Eff Clin Pract 2000;3:277–283.

Steel K, Gertman PM, Crescenzi C, Anderson J. Iatrogenic illness on a general medical service at a university hospital. N Engl J Med 1981;304:638–642.

Thomas EJ, Brennan TA. Incidence and types of preventable adverse events in elderly patients: population based review of medical records. BMJ 2000;320:741–744.

Thomas EJ, Studdert DM, Burstin HR, et al. Incidence and types of adverse events and negligent care in Utah and Colorado. Med Care 2000;38:261–271.

Thomas EJ, Studdert DM, Newhouse JP, et al. Costs of medical injuries in Utah and Colorado. Inquiry 1999;36:255–264.

Villanueva EV, Anderson JN. Estimates of complications of medical care in the adult US population. BMC Health Serv Res 2001;1:2–10.

Vincent CA. Research into medical accidents: a case of negligence? BMJ 1989;299:1150–1153.

Vincent CA, Neale G, Woloshynowych M. Adverse events in British hospitals: preliminary retrospective record review. BMJ 2001;322:517–519.

Wilson RM, Runciman WB, Gibberd RW, Harrison BT, Newby L, Hamilton JD. The quality in Australian health care study. Med J Aust 1995;163:458–471.

# 2 Perceptions of medical error and adverse events

In chapter 1, we defined medical error. Definitions, however, do not necessarily represent how something is interpreted in practice. Beyond postulating what medical error is and how to prevent it, researchers and, eventually, policymakers have to examine how the people who are exposed to medical error interpret it if they want to effectively understand it. All individuals involved in the health service process have different ways of perceiving medical error. This means that physicians, nurses, patients and potential patients, and students and residents all have the potential to interpret medical error in ways other than according to the definition postulated by theorists in the medical field, and these variations in perception can lead them to act in ways that are radically different from what might be expected.

## 2.1 Introduction

Even if perceptions of medical error differ, one may wonder why it is important to understand how perceptions differ among participants in the health care system. One might think that a medical error is a medical error, no matter how it is perceived. This, however, is not strictly true. How people perceive something and what they believe to be true impacts how they interact with, apply, and are affected by something. No matter how theorists define medical error, perceptions play a critical role in creating its "reality," so, in practice, the characteristics and impacts of medical error are shaped by how medical error is perceived. How medical error is perceived, in turn, is constantly being redefined by what is identified as an error and how it is explained in the medical field. For this reason, perception of medical error can determine the efficacy of initiatives that are implemented to improve patient safety. Bosk (2005) suggests that the IOM's report completely ignores a certain social-science approach to analyzing error, one that looks how workers define error "on the shop floor." Bosk argues that the strategy of understanding perceptions is absent from current research; without it, no policymaker can hope to reduce the frequency of medical error. By understanding the ways in which perceptions of medical error can differ, researchers and policymakers can become more aware and thus create better policy.

Another reason why it is important to understand perceptions of medical error is that doing so can help researchers develop an awareness of the barriers to building a culture of safety, where the minimization of medical errors is the top priority. Interpreting the perspectives of front-line workers with direct but different experiences can help to pinpoint problems and possible solutions to them that might otherwise go unremarked. As Edwards et al. (2008), citing Nieva and Sorra (2003), point out, "assessing staff perceptions of safety culture is a first step to identify areas in need of improvement, and then, follow-up improvements can show the effectiveness of changes made to address initial improvement needs." Furthermore, knowing how perceptions of medical error differ among physicians, nurses, and medical students can help researchers to understand the context in which safety initiatives are being implemented and the mindset of the people who will be performing them. Moreover, because perceptions are not static, Edwards et al. (2008) conclude that researchers can effectively understand the efficacy of different safety initiatives in terms of moving medical environments toward safety cultures by analyzing changes in perceptions over time. The importance of this move is highlighted by recently published studies that show that physicians are still uncomfortable with reporting medical errors, even though a large part of the overall push in the medical community toward improved patient safety has been the establishment of nonpunitive reporting systems (Anderson et al., 2009). Without having effective reporting systems in place, it is impossible to accurately measure and address medical error. Therefore, researchers need to have a detailed understanding of the effectiveness of patient safety initiatives, rather than just relying on medical error statistics. Measuring efficacy in terms of the propagation of a safety culture, which builds on and develops the perceptions and, eventually, the practices of medical workers, provides researchers with more nuanced information than they would have if they were just to measure the efficacy of different safety initiatives in terms of a strict reduction in medical errors.

If researchers want to truly understand medical error, they must first have an understanding of the different ways in which medical error is perceived by those who, through their actions, are constantly redefining the part medical error plays in the medical and health care field. If policymakers want to put into place effective policies for combating medical error, they must first have an understanding of the different ways in which medical error is perceived by those who will enact and be affected by those policies. Essentially, by striving to understand the perception of those who must effectively negotiate the reality of medical error, they are doing no less than enhancing their own perception.

## 2.2 Perceptions by physicians

As indicated in chapter 1, in the section on medical error, definitions of medical error can vary. How a physician defines medical error can change how he perceives his own performance, how he perceives his colleagues' performance, and, as a result, what he chooses to report. Even physicians who work together can have different ideas of what constitutes an error (Elder et al., 2004). Elder et al. (2004) conducted research in which they attempted to show how a physician comes to a certain perception of medical error. They first collected definitions of medical error from the vast amount of literature that has already been written on the subject. Next, they surveyed family physicians, presenting them with hypothetical scenarios and asking them if they felt a medical error had taken place. Finally, they compared the various definitions of medical error and their survey results to see if they could discover a trend to whether or not physicians perceived something as a medical error.

Elder et al. (2004) found that there are three primary things that a physician looks for when deciding whether or not an error has occurred. First, she ascertains whether or not there was harm from the event; second, she determines if the event is a rare or common occurrence; third, she judges whether she considers an individual or the system to be at fault for the event. If an event is considered to be harmful and rare and the result the actions of an individual, then physicians are more likely to consider it an error (Elder et al., 2004).

The findings of Elder et al. (2004) draw attention to three discrepancies between the generally accepted definitions of medical error and the way in which physicians define an error. The Institute of Medicine (IOM) definition identifies "errors of execution" and "errors of planning" (Kohn, 2000). Neither of these types of error depends on outcome; each is defined solely on the basis of what action was taken. If a physician perceives only those medical events that are harmful as error, she is unlikely to also recognize actions that could be potentially harmful in the future. The IOM definition of medical error also potentially differs from the perception of physicians in that physicians are more likely to define those events that are rare occurrences as an error, whereas the IOM does not distinguish between rare and common occurrences. Furthermore, physicians are more likely to consider an event as an error if they perceive it to be the responsibility of an individual, whereas the IOM draws attention to the fact that errors are most commonly perpetuated through systemic causes. These two things may be related in that errors caused by the system are generally less recognizable and therefore may be seen as common, unavoidable occurrence, whereas individual errors are easier to identify and therefore stand out when they do occur.

The danger in these discrepancies is that an error, whether or not it is harmful, whether it is common or rare, and whether it results from individual or systemic cause, ultimately leads to suboptimal health delivery at the very least; at the most, it can lead to the occurrence of large serious errors. This notion is explored further in chapter 3. Furthermore, given that physician reporting is the foundation of many medical error prevention initiatives, it is important that the physician perception of medical error match more complex formulations of medical error. While a physician might have difficulty in addressing systemic errors alone, the effectiveness of a larger task force in addressing these systemic errors depends on the physician's ability and willingness to recognize and report those errors that come from the system (Hobgood et al., 2005).

## 2.3 Perceptions by the public

Burroughs et al. (2007) suggest that the patient definition of what constitutes a medical error is broader than the standard understanding of medical error by medical practitioners (as cited in Hospital Peer Review, 2007). They find that communication problems between practitioners and patients, practitioners' failure to respond in a way that patients see as appropriate, and patient falls in health care settings are all considered medical error by patients and yet are not categorized under the traditional, or medical system, definition of medical error. Such studies draw attention to the wide differences in perception of what constitutes medical error. Mazor et al. (2010) note that in many cases where parents perceive an error to have taken place in the case of their child, that feeling arises because they do not believe that their physician has taken them seriously; in other words, they perceive the care provider as being inadequately responsive, which is defined by the medical system as a service issue. As Northcott et al. (2008) put it, patients define medical error as including "problems with service quality."

Even if no medical system-defined error has occurred and the care provider has done what is needed to prevent physical harm, a perceived medical error can result in emotional and psychological harm to the patient and his family (Mazor et al., 2010). Therefore, whether or not a medical system-defined error has actually taken place, perceived medical error is an important part of determining how effective or satisfactory medical care has been. It follows, then, that educating patients and their families as to what to expect during their experience with medical services – whether or not they experience an actual medical system-defined error – can help prevent unsatisfactory medical experiences. Further, educating patients about how to observe their own bodies can help to lessen medical error by helping patients become more part of the monitoring and reporting team. This may have the side benefit of making patients feel more responsible and in control of their medical experience, thus, increasing patient satisfaction.

The patient perspective demonstrates how understanding differing perspectives of medical error must be developed if all parties involved in health services are to work together to improve patient safety. Even patients have a role to play in lessening the occurrence and impact of medical error. A shared understanding of what constitutes medical error can be built through, as Burroughs et al. (2007) specifically recommend, patient education programs "to address the fears and concerns of each patient." It is crucial, however, to understand why patients' perceptions of medical error are so different from those of medical practitioners.

Public perception of the safety of the medical system, as influenced by media, friends and family, and experiences, can lead patients who come to the hospital to expect to face a medical error. Even if no error takes place during a patient's medical experience, if he perceives an error to have taken place, his overall satisfaction with the experience will be altered. Some researchers argue, however, that, rather than including everything the patient considers to be medical error in a standard definition of medical error, medical caregivers must search for the root cause of why patients are so broadly defining medical error. Patti Muller-Smith, in an interview with Hospital Peer Review, draws attention to why this is important when she says that it would be a mistake to expand the definition of medical error to fit that of patients (Hospital Peer Review, 2007). She contends that doing so would obscure the focus on making changes that would prevent significant medical errors, particularly those that have an impact on patient safety. Still, it is hard simply to maintain that tight focus on medical system-defined error without understanding patients' perception of their experience, as this is dependent upon communication. Postmedical-experience surveys of patients already assist in measuring whether patients feel that their needs and concerns were adequately addressed. It is more difficult, however, to know whether patients feel that they are part of a satisfactory communication process. Even if the patient does not realize it, she might not be receiving the communication necessary to understand what exactly to expect of her hospital experience.

The solution, then, is to work to ensure the strength and effectiveness of patients' communication with their caregivers; in particular, patients need to be educated so that their expectations of the medical process match those of their caregivers, enabling them to comprehend more clearly the medical system understandings of medical error. Furthermore, there must be processes during a patient's medical experience through which they can communicate whether they feel they are being adequately cared for and, if not, why they feel unsafe. This way, health service providers can work to address widespread sources of patient anxiety and increase the chances that patients will feel satisfied after the experience.

Ultimately, the best way to ensure that patients are not anxious about the possibility that a medical error will occur is to lower the occurrence of medical errors. Previous experience with medical error, from either personal experience or from exposure to the media, can intensify their expectation and assessment of error. Coverage of the most extreme medical errors in the media, for example, can lead patients who come into the hospital to expect to face an error. If patients expect an error, they are more likely to see an error in anything that seems abnormal to them.

Several researchers have shown how previous exposure to medical error can result in patient and public anxiety. Northcott et al. (2008) report that those people who believe that either they themselves or a family member has experienced a preventable medical error are more likely to overestimate the frequency with which medical errors occur, have less trust that their doctor would report a medical error, and have a lower opinion of the health care system overall. Mazor et al. (2010) found that perceived medical errors have strong impacts on patients and their families, and Mira et al. (2009) conclude that "risk perception for adverse events increases after having suffered such an event."

There is the potential, then, for a somewhat self-perpetuating cycle in which, as patients become increasingly anxious about the medical system, they are more likely to perceive a situation as being part of a medical error, therefore feeding their anxiety. This cycle might not be as vicious if patients were shown the gravity with which the medical field treats medical error. Mazor et al. (2010) found that, in the event of an error, the involved medical practitioner's acknowledgment of that error often tempered the patient and family's subsequent view of the medical system, especially if it was accompanied by an apology. Patient and public perception shows that those in medical services not only must work to reduce actual medical errors but also must work to rebuild the public image of their field so that their patients once again recognize it as consistently and reliably safe.

Besides addressing the impact of perceived medical error, respecting patient and family perspectives could actually assist in avoiding those medical errors that fit theorists' definitions. Mazor et al. (2010) suggest that parents and other family members might have a better understanding of the patient and might even help prevent medical errors, for example, by questioning misdiagnoses. By regarding patient perception as valuable and worthy of consideration, medical practitioners, policymakers, and theorists can all take steps to lessen the instances and effects of error.

## 2.4 Perceptions by health care staff

Physicians are not the only medical professionals. Others, including therapists, paramedics, and nurses, can all play a part in perpetuating or preventing medical error. In particular, many studies have focused on nurses and their unique perspective on medical error, including, for example, how it affects their reporting practices and how it develops from their relationship to physicians.

Nurses perform many of the frontline tasks in medical settings; they are often responsible for the majority of patient interaction (Ramanujam et al., 2008). In particular, Ramanujam et al. point out that "nurses are responsible for the administration of medications, assessment of patients' condition, supervision of patients' activity, [and] oversight of unlicensed personnel" (2008, citing Institute of Medicine, 2004; Ramsey, 2005). Given their high level of exposure to patients, nurses have the potential to play an important role in the identification of medical errors and the promotion of a culture of safety. Promoting a culture of safety is seen as one of the most effective ways of preventing medical error. A culture of safety depends on a group of professionals who, both individually and as a group, are willing to dedicate themselves to prioritizing the prevention of medical error as part of their pursuit of improved patient safety. Part of this group dedication includes the ability and the willingness to learn from each other. Nurses often work alongside other medical professionals, particularly physicians, and the cooperative and supplementary nature of their work puts them in an ideal position to observe and provide feedback for those other professionals working with them.

The ideal advantage that nurses may enjoy in terms of preventing medical error may conflict with the real pressures nurses face in their work. As Elder et al. (2008) discovered, for example, nurses perceive that there are often too many barriers for them to have a useful function in reducing medical error. Despite the fact that, as described earlier, medical error reduction is often promoted as a product of teamwork, nurses often feel that they are at risk of straining their relationships with their coworkers if they are discovered to have reported a medical error (Elder et al. 2008).

These barriers to medical error reporting mean that a nurse might not report a medical error if no harm actually befalls the patient (Elder et al. 2008). One can see a similarity between a nurse's reluctance to report a medical error that does not result in patient harm and the way in which physicians decide whether or not an event constitutes a medical error on the basis of whether or not harm occurs (Elder et al. 2004). Harm occurrence, as a limiting factor to whether or not a nurse perceives medical error as worth the effort and the cost to work relationships that would come from her reporting it, means that medical errors that are part of the system are unlikely to be identified until they do result in serious harm. Even though medical error reporting has a potential to be proactive, this limitation often relegates medical error to the purely reactive.

Besides recognizing and reporting errors made by those with whom they work, nurses also have to recognize and decide when to report their own errors. One can see how self-reporting could be a stressful process to go through. Elder et al. (2008) describes nurses who see making an error as a deeply shameful thing to do. However, they also describe nurses who hold a great deal of trust and respect for those fellow nurses, whom they see as being honest and trustworthy enough to report their own errors.

The latter perspective especially makes sense in the context of medical error, given that the majority of medical errors have been attributed to systemic causes. Self-reporting should not be seen as shameful for two reasons. First, if medical errors are caused by a systemic flaw, then the vast majority of time an error cannot be attributed to an individual. If a medical error should bring shame on anyone, more often than not that shame should accrue to the entire system. The second reason that self-reporting should not be shameful is that doing so shows a prioritization of the success of the entire medical system, since a report of an error associated with a particular person can assist in the identification of systemic problems. Both of these reasons can also apply to why a nurse's colleague should not be embarrassed or angry if that nurse reports one of his errors; that error likely is not the individual's fault, and identifying that error is a boon that will help all medical professionals in the system do their jobs better.

If nurses feel uncomfortable pointing out errors directly to a peer, an anonymous reporting system can assist in avoiding this discomfort. On the other hand, sometimes even an anonymous report can be easily traced back to a single nurse, in cases, for example, where that nurse was the only other person present when the error took place. In these cases, it is important to have not only an anonymous reporting system but also a culture in which the nurses share a perspective that medical error reporting is a laudatory process in which to participate. Such a perspective becomes even more important in situations where nurses identify the errors of those who they might feel are higher in the hierarchy.

Despite the fact that nurses often work closely with physicians and residents, Elder et al. (2008) found that nurses rarely report a physician's or resident's medical error. Nurses describe having to find ways to draw physician's attention to errors without embarrassing them by forthrightly pointing them out. Again, and again problematically, nurses found it worth the risk of a negative reaction from a physician only if they saw that a patient was or had been immediately in harm's way. While there is a well-documented hierarchal tension between nurses and physicians, one hopes that it can be overcome by awareness that medical error should be considered irrelevant to status struggles between individuals; it is the product of a faulty system.

In the section of this chapter on physicians, we suggested that physicians are less likely to identify medical error that stems from systemic causes than that which results from individual error. Nurses, however, have been shown to recognize the relationship between organization and patient safety and therefore have the potential to identify systemic causes of medical error. Ramanujam et al. (2008) conclude that nurses' perceptions of patient safety conditions in their workplaces change as their work demands change. As the demands on nurses increase, their perceptions of how safe their workplaces are go down. Nurses see higher work demand as coming from work overload, coupled with exhaustion and the inability to have personalized relationships with patients. Higher work demand is perceived by nurses as correlated with decreased patient safety (Ramanujam et al. 2008). Given that exhaustion and overload are two commonly identified issues in workplace organization and structure (see chapter 3), it appears that nurses' perceptions could be valuable in flagging certain systemic circumstances that could lead to medical error.

As Edwards et al. (2008), citing Ralston, et al. (2005), point out, a medical error is often only apparent to those who are providing the care during which the event occurs, and whether or not the error is reported depends on the people who committed it.

## 2.5 Perceptions by medical students

Current medical students are in a unique position with regard to perceptions of medical error. Students of the past decade are the first to be educated after the release of *To Err Is Human*, the Institute of Medicine's landmark report on medical error. The IOM report brought global attention to an issue that, while previously known about and studied, had not been a major focus of the field. Now, medical error and patient safety are increasingly incorporated into medical curricula. Patient safety groups can research student perceptions to gauge the effectiveness of efforts to inculcate a better understanding of how to prevent medical error in the next generation of medical practitioners. Having been educated only in the context of an education system that is increasingly aware of the dangers of medical error, recent medical students are also likely freer of old biases and may well be more likely to eventually innovate new creative solutions to patient safety problems. Medical students represent the leading edge of the medical field. A change in medical students' perception of error would be indicative of the beginning of a shift in perception in the larger field.

Even as early as 2003, Schenkel et al. showed that residents have some understanding of what constitutes medical error. In their study, residents in the emergency department attributed responsibility for medical error primarily to the overall emergency department team (Schenkel et al., 2003). While they also identified the actions of individuals as a contributing factor, their primary focus on the entire department displays an understanding of the systemic nature of error. Schenkel et al. (2003) noted that residents were more likely to recognize diagnostic errors, which they suggest might be a result of the educational focus on differential diagnosis, a technique where physicians make a list of diagnoses and then eliminate diagnoses from the list until one remains. If none remain, it suggests that a misdiagnosis may have occurred. In other words, residents may have had limited awareness of medical error because of standard curricula, but the residents did not yet have an indication that patient safety–based education had been effective.

Medical education is the most obvious way students' perceptions of medical error are shaped. Muller and Ornstein (2007) found that students with clinical training were more likely to say that their views of medical error had changed over time than were students without such training, though it is unclear how these perceptions had changed. What this indicates is that educators have to be prepared to adapt the way in which they teach students to integrate a certain view of medical error into their perspective depending on whether they are in the classroom or in the field.

At the same time, it is insufficient to regard medical education as the sole variable affecting student perspective on medical error. While medical education formally stresses the importance of error reporting, among other patient safety initiatives, students, like physicians and nurses, are subject to a number of contradictory pressures. In particular, students are participating in a field that is notoriously competitive, where those who can withstand the demands of the field and manage to succeed in delivering quality care stand to gain valued recognition, whether it is in the form of a better residency or better job placement, respect from their peers, or, potentially, a higher salary. Under certain conditions, it might be more convenient for students to deflect blame for error that is of uncertain origin. Students might attempt to capitalize on the focus in the field on systemic error and use it as an opportunity to take advantage of plausible deniability, shifting attention away from their own potential responsibility. What is more likely is that in most circumstances there are multiple causes for error; in different cases, one or another might be given more weight, but generally all have some degree of effect.

By studying medical students, researchers can examine variables that affect physicians and other medical staff in some of their most intense forms. Given that one of the primary causes of medical error is fatigue (see chapter 3), those concerned with patient safety look to work-hour overload as a source of increased error. When it comes to potential fatigue due to increased work hours, residents represent the extreme in the medical field. In an effort to give them the maximum opportunity for hands-on training possible, medical students in residence are often expected to work intense hours, sometimes more than 80 hours a week. In 2003, the American Accreditation Council for Graduate Medical Education (ACGME) set a limit on the number of hours that all residents could work: a maximum of 80 hours total a week and a maximum of 30 hours in continuous shifts (Philibert, 2002, cited in Jagsi et al., 2008). Some researchers, however, have expressed concern that the effectiveness of these limited hours in reducing medical error will be counteracted by increased patient workload (Okie, 2007, cited in Jagsi et al., 2008).

Jagsi et al. (2008) investigated the effect of reduced hours on residents' perceptions of patient safety. They sent surveys to residents in all ACGME-accredited training programs at two hospitals in Boston, Massachusetts, between 2003 and 2004 and obtained 1,770 responses across the spectrum of specialties (Jagsi et al., 2008). What they found was that, between 2003 and 2004, when the ACGME's work-hour limits were set, a significantly lower number of residents reported committing medical errors, particularly those caused by "working too many hours" and "carrying or admitting too many patients" (Jagsi et al., 2008). Students whose hours were reduced were also less likely to identify fatigue as a factor that frequently affected the quality of care they provided (Jagsi et al., 2008). Their findings show that not only did those residents whose hours were reduced not have increased workloads, but also they reported fewer medical errors. The work of Jagsi et al. (2008) is an example of using the perceptions of medical practitioners as a way of measuring changes to patient safety. Their findings indicate a degree of success in identifying patient safety initiatives that can be used to guide further research and policy for others in the medical field.

It is difficult, however, to identify a single variable that affects medical error and then to generalize it to the broader field of patient safety. Jagsi et al. (2008) also found that one reason that residents' patient loads were not increased when their hours were limited is that those hospitals that had to decrease their residents' hours also took steps to ensure that patient overload would not become a problem, such as taking on additional residents and physicians. That these hospitals had the foresight to do this is an example of making organizational adjustments in anticipation of new opportunities for systemic errors. It illustrates the importance of considering the entirety of the system when making changes.

What further problematizes the search for particular identifiable perspectives on medical error is that the students' perspectives cannot be distilled into one single position. While there might be observable trends across all students, given that all students follow similar educational paths, certain core demographic differences can alter perceptions of medical error. While Muller and Ornstein (2007) found that there is no significant difference in how male and female students define medical errors or in their willingness to report errors, they reported that female students are more likely to experience feelings of guilt after committing a medical error and more likely to feel stress about potential consequences.

## 2.6 A sociological perception of medical error

Despite the large body of literature that further corroborates and expands on the analysis of medical error featured in *To Err Is Human,* one would be remiss if one failed to mention that there exist academic perspectives that do not support the IOM's description of medical error. At the beginning of the chapter, we cited Charles Bosk, who felt that the IOM report ignored one way of understanding medical error. In 2005, Bosk, who wrote about medical error as early as 1979 in his book *Forgive and Remember: Managing Medical Failure,* published a paper called "Continuity and change in the study of medical error: The culture of safety on the shop floor." While his article (2005) does not disagree with the need for a culture of safety and even begins by praising the 1999 IOM report for its role in bringing public attention to the issue of medical error, Bosk does believe that the report focuses on one way of understanding error while ignoring another.

Bosk's description of how the IOM interprets medical error should be familiar after reading chapter 1. He says that IOM uses "normal accident theory, a blend of organizational theory, cognitive psychology, and human factors engineering" to explain medical error (Bosk, 2005). In normal accident theory, systems are "error-prone," accidents are "normal," and individuals do not fail but instead are the victims of inevitable failings in the system. "Normal accidents" occur because each part of the system is individually complex while also having an effect on the complex working of the overall system; small errors therefore have consequences for the whole system (Bosk, 2005).

In "normal accident theory," error reduction occurs through better system design. Bosk (2005), however, says that the IOM completely ignores a second social-science approach to analyzing error, one that looks to understand the workers' perceptions of error. The IOM, therefore, is "unmindful of the cultural context of the workplace" (Bosk, 2005). In other words, the IOM fails to propose solutions that deal with the cultural realities of the field. Through his work, Bosk is actually calling on researchers, theorists, policymakers, and, in particular, the Institute of Medicine to change their perceptions.

## 2.7 Conclusion

Medical error does not exist outside society. To understand medical error, one must know not just how it is caused but also how it perceived, especially by those who have an effect on the prevalence and prevention of error, as well as by those who are affected by it. Perceptions can help explain why certain patient safety initiatives are or are not effective, why patient satisfaction is not directly linked to how the health system rates patient safety, and even how health researchers interpret and apply their findings. It is not enough to put in place systems for reporting errors without understanding that all those involved in medical processes have different views on what constitutes an error. If the IOM simply calls for policy to address error that is systems based, they are bound to be ineffective, because they fail to address the fact that, in practice, nurses and physicians tend to attribute errors to individuals. It is insufficient for researchers and policymakers to study how or how much medical error occurs. For their efforts to be relevant and valuable, they must study how medical error is perceived. It is a difficult task, but a necessary one.

### References

Anderson B, Stumpf PG, Schulkin J. Medical error reporting, patient safety and the physician. Journal of Patient Safety 2009;5:176–179.

Bosk, CL, Continuity and Change in the Study of Medical Error: The Culture of Safety on the Shop Floor. Princeton, NJ: Institute of Advanced Study, School of Social Science, University of Pennsylvania, School of Social Science. Occasional Papers 2005; Paper 20.

Bosk C. Forgive and Remember: Managing Medical Failure. Chicago: University of Chicago Press; 1979.

Burroughs TE, Waterman AD, Gallagher TH, et al. Patients' concerns about medical errors during hospitalization. Jt Comm J Qual Patient Safety 2007;33:5–14.

Edwards PJ, Scott T, Richardson P, et al. Using staff perceptions on patient safety as a tool for improving safety culture in a pediatric hospital system. Journal of Patient Safety 2008;4(2):113–118.

Elder NC, Vonder Meulen MB, Cassedy A. The identification of medical errors by family physicians during outpatient visits. Annals of Fam Med 2004;2(2):125–129.

Elder NC, Brungs SM, Nagy M, Kudel I, Render ML. Nurses' perceptions of error communication and reporting in the intensive care unit. Journal of Patient Safety 2008;4(3):162–168.

Hobgood C, Eaton J, Weiner BJ. Identifying medical errors: developing consensus on classifications and consequences. J Patient Safety 2005;1(3):138–144.

Hospital Peer Review. Patients may define medical errors differently than you. Hospital Peer Review 2007;32(4):41–42, 47–48.

Institute of Medicine. Keeping Patients Safe. Transforming the Work Environment of Nurses. Washington, DC: National Academies Press; 2004.

Jagsi R, Weinstein DF, Shapiro J, Kitch BT, Dorer D, Weissman JS. The Accreditation Council for Graduate Medical Education's limits on residents' work hours and patient safety. Arch Intern Med 2008;168(5):493–500.

Kohn LT, Corrigan JM, Donaldson MS, editors. Committee on Quality of Healthcare in America, Institute of Medicine. To Err Is Human: Building a safer Health System. Washington, DC: National Academy Press; 2000.

Mazor KM, Goff SL, Dodd KS, Velten SJ, Walsh KE. Parents' perceptions of medical errors. J Patient Safety 2010;6(2):102–107.

Mira J, Lorenzo S, Vitaller J, et al. Hospital clinical safety from the patient's point of view. validation of a safety perception questionnaire. Rev Med Chile 2009;137:1441–1448.

Muller D, Ornstein K. Perceptions of and attitudes towards medical error among medical trainees. Medical Education 2007;41:645–652.

Nieva VF, Sorra J. Safety culture assessment: a tool for improving patient safety in healthcare organizations. Qual Saf Health Care 2003;12: ii17–ii23.

Northcott H, Vanderheyden L, Northcott J, et al. Perception of preventable medical error in Alberta, Canada. Int J Qual Health Care 2008;20(2):115–122.

Okie S. An elusive balance: residents' work hours and the continuity of care. N Engl J Med 2007;356(26):2665–2667.

Philibert I, Friedmann P, Williams WT. New requirements for resident duty hours. JAMA 2002;288(9):1112–1114.

Ralston J, Larson E. Crossing to safety: transforming healthcare organizations for patient safety. J Postgrad Med 2005;51(1):61–67.

Ramanujam R, Abrahamson K, Anderson JG. Influence of workplace demands on nurses' perception of patient safety 2008;10:144–150.

Ramsey G. Nurses, medical errors, and the culture of blame. Hastings Rep 2005;35:20–21.

Schenkel SM, Khare RK, Rosenthal MM, Sutcliffe KM, Lewton EL. Resident perceptions of medical errors in the emergency department. Acad Emerg Med 2003;10(12):1318–1324.

# 3 Causes of medical error and adverse events

The complexity of the health care delivery system may be primarily responsible for increased error rates. A large proportion of the blame lies not with individual failures but with system complexities. The second IOM report, *Crossing the Quality Chasm: A New Health System for the 20th Century,* states that safety is a systems property and that patients should be safe from injury caused by the care system (2001). The report suggests that better reporting of systemic errors will lead to better system design, minimizing latent system defects and, in turn, medical errors. The report goes on to elaborate that reducing risk and ensuring safety require greater attention to systems that help prevent and mitigate errors. It is suggested that little can be achieved in improvements to patient safety unless safer systems are designed around human factors. Health care systems should be held accountable, rather than just individuals. The health care system requires coordinated improvements to overcome problems in system design that threaten quality of care.

## 3.1 Introduction

It has long been recognized that health care systems are far from safe. To offset the sources of these deficiencies, health care facilities have employed safety programs since the beginning of the 20th century. Physicians created the initial model for external quality oversight in the United States in 1917 (Roberts et al., 1987). This was soon followed by the establishment of the hospital standardization program of the American College of Surgeons, which was the forerunner in the United States of both the National Joint Commission on Accreditation of Health Care Organizations (JCAHO), established in 1951, and the federal and state regulatory models now in place for all types of health care organizations. The commitment to quality and patient safety by JCAHO has been periodically strengthened ever since.

The foremost requirement of designing systems in a way that can eventually reduce the possibility of errors is to allow the identification of errors before significant damage occurs by integrating disciplines, ideas, and various models and theories. Health care has liberally borrowed concepts from many sectors, particularly industry and social psychology. Creating safer systems requires more such knowledge transfers with modifications to suit our needs. It is our belief that for quality improvements and patient safety initiatives to occur, we will need to absorb various ideas and theories from human factors reengineering, information technology, organizational and management sciences, cognitive psychology, and bioethics.

Research has described two kinds of errors, active and latent (Rasmussen and Pedersen, 1984). Active errors are those that produce immediate events and involve operators of complex systems; examples are errors made by pilots in aviation and by health care professionals in medicine. Latent errors may result from factors that are inherent in the system. These include organizational issues, such as excessive workload, insufficient training, and inadequate maintenance of equipment. Latent errors lie dormant in the system, waiting to strike at the opportune time. These latent errors play a critical role and influence the operators to commit the final active error. This concept has been well summarized by Reason (1990). Reason (1990) noted, about active errors: "their part is usually that of adding the final garnish to a lethal brew whose ingredients have already been long in the cooking." Another suitable analogy for this concept has been proposed that sees latent errors and active errors as resembling the base and the sharp end of a pyramid, respectively (Cook and Woods, 1994). Organizational factors, such as the hierarchical structure, management policies, and the work environment, including workload, may form the blunt end or the base of the pyramid (see ▶Fig. 3.1). Health care professionals operate at the sharp end of the pyramid, and their deficiencies are manifested in the form of mistakes, violations, and incompetence. The organizational factors at the base play a critical role in influencing these individual deficiencies and ultimately cause the error. Unarguably, the sharp end alone is a component of the

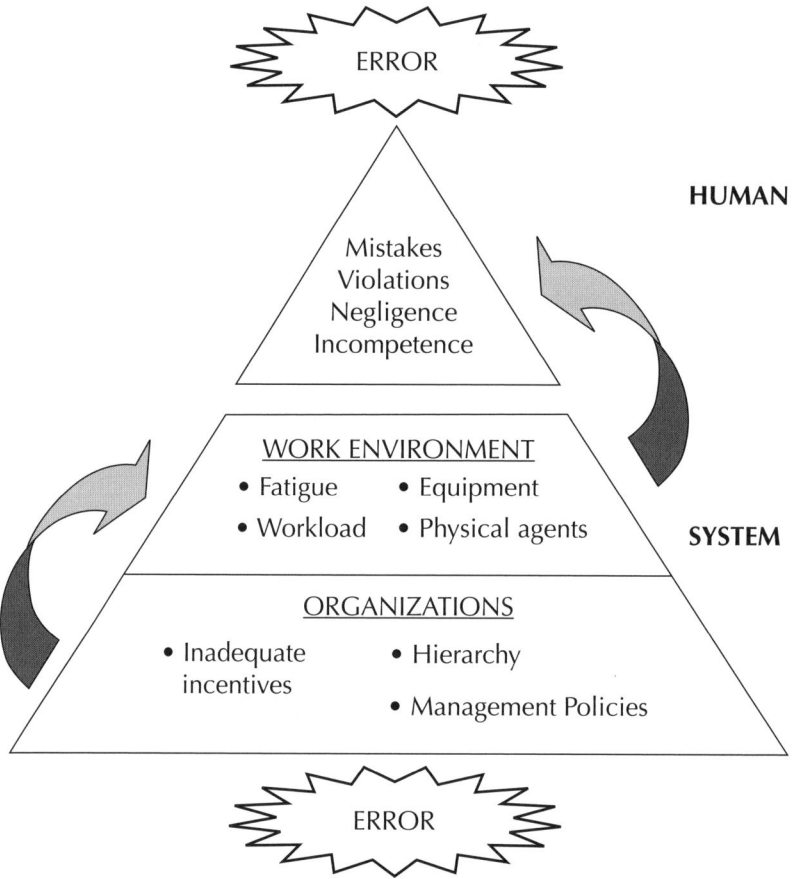

**Fig. 3.1:** Influence of system factors on human deficiencies in causing an error (from Cook and Woods, 1994, modified).

system that represents only a small part of the overall problem; therefore, measures aimed at this end can be minimally effective. Maximum efforts have to be directed at the latent deficiencies in the system, the base of the pyramid, which will ultimately yield more positive results than can be obtained by minimizing the active errors at the sharp end. It is generally agreed that the most immediate and proximate error appears to be human error, while the background and provoking error are well beyond an individual's control (Leape, 1994).

The conventional methodology of error investigations, in which, after a careful retrospective analysis, errors are primarily classified as human failure and the majority of blame is placed on either human error or mechanical error, must change. These investigations lead to such shortsighted measures as revising procedures, taking administrative and regulatory actions, and introducing new rules that add to the complexity of the process. Thus, increased complexity serves to introduce new types of failures that present new opportunities for error generating behavior (see ▶ Fig. 3.2). It is, therefore, suggested that health care errors be viewed from a systems perspective to help bring about effective changes in delivering safe health care.

Attitude and stress are two important system factors that play a major role in influencing the active errors and that deserve further elaboration. A certain part of responsibility for the high incidence of errors in health care may be attributed to the attitude of health care professionals toward error. It has been suggested that physicians tend to overestimate their ability to function under conditions of stress, such as fatigue and high anxiety (Helmreich, 2000). This belief among health care professionals that they

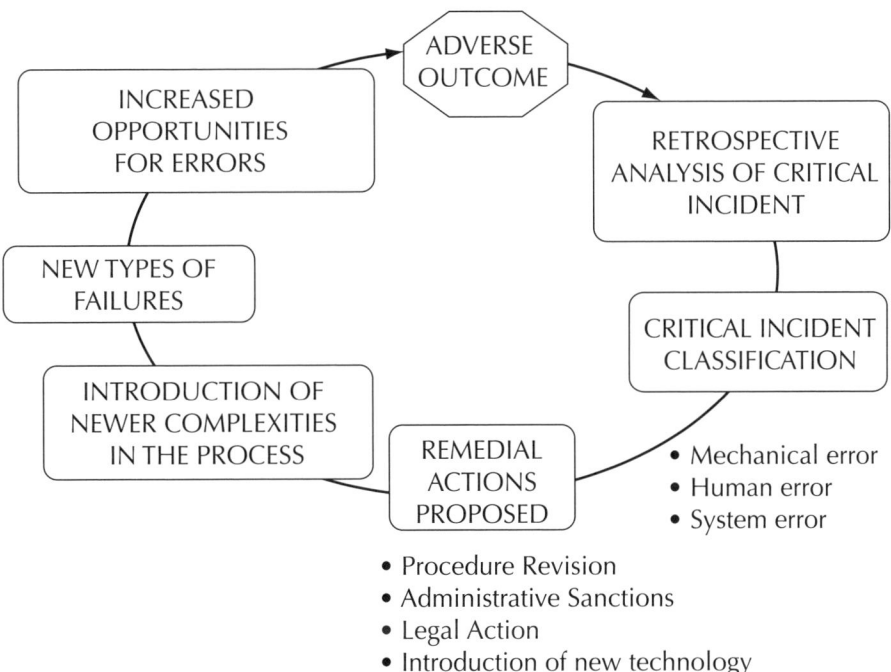

**Fig. 3.2:** Cyclical propagation of error-generating behavior.

can perform flawlessly even at times of extreme stress may prove to be their undoing. Health care professionals are only human and are as susceptible to error as any others, but there seems to be a prevailing consensus that perfection is their forte and mistakes are unacceptable. This perception seems to be prevalent not only among the general public but also within the health care fraternity. It is therefore appropriate that in his editorial remarks, Blumenthal (1994) calls for a total attitude change toward errors from both physicians and the general public. Blumenthal also suggests that such efforts require abandoning traditional cherished myths. Health care, just like any other human-controlled process, is not 100% safe. It is rather a very risky business, simply because it is not delivered by a machine: a majority of health care processes are run by, controlled by, and dependent on human function. It is appropriate for both health care providers and health care seekers to accept this reality.

It has been more than a half-century since it was reported that stress directly influences an individual's thought processes and that, as stress increases, an individual's thought processes and attention span narrow (Combs and Taylor, 1952; Easterbrook, 1959). The stressful conditions can due to mental or physical factors. Stress and fatigue have recently been documented to be the principal contributors to many accidents (Dinges et al., 1997; Sandal, 1998). These authors also suggest that, under stressful conditions, even well-rehearsed actions are poorly performed. However, medical staffs are more likely to deny the effects of stress and fatigue on their performance (Sexton et al., 2000).

Sleep deprivation has been identified as one of the main concerns in the practice of medicine. Long, continuous working hours and sleep deprivation are incompatible with enhancing quality care and patient safety (Shine, 2002). Medical trainees, who share a great part of the health care burden, have an immense workload and suffer from the accompanying sleep deprivation and fatigue (Schwartz et al., 1992). Sleep deprivation is considered the prime cause of many health care errors committed by trainees (Gaba et al., 1994). With fatigue and associated stress topping the concerns of the sleep-deprived health care professional, provision of quality care and patient safety may not be a priority. It is again the system at play. The transportation and aviation industries have long recognized the disastrous consequences of fatigue for safety and have implemented strict regulations (Hopkins, 1971; Office of Technology Assessment, 1991).

Health care is the only high-risk setting that may have overlooked the fatigue factor and therefore failed to implement any regulations limiting professional staff's working hours. Resident physicians seem to be suffering the most from these unregulated work hours. In July 2003, the Accreditation Council for Graduate Medical Education (ACGME) in the United States mandated an 80-hour workweek for all medical residents and fellows in accredited training programs. Resident physicians provide a source of expertise at less cost than that associated with more experienced doctors; in addition, the training years are crucial for honing residents' clinical skills. It is therefore believed that the more time residents spend training, the better the skills that they acquire – but, in the meantime, patient safety may suffer.

Concerns have also been raised that the recent work-hour restrictions imposed by the ACGME have undercut the importance of continuity of care in surgical practice (Fischer, 2004). This argument is validated by the finding that a substantial number of residents reported negative impacts on surgical training and quality and continuity of patient care due to work-hour limitations in New York State (Whang et al., 2003). In another, similar survey, fewer than half of residents expected the work-hour reforms to have a positive impact on patient care. Romanchuk (2004), however, suggested that the new regulations would result in a happier work setting that would facilitate improved patient care. The impact of the changes should be continuously monitored over time before researchers reach any conclusions. It is also interesting to note that the new work-hour limits imposed by the ACGME exceed the allowable number of work hours in many other hazardous industries. Also, an arbitrarily selected 80-hour workweek may not necessarily reduce the occurrence of medical errors by trainees. In this context, the findings of Davydov et al. (2004) are worth consideration. These authors reported that they found no correlation between the number of hours worked and the number of prescribing errors made by medical house staff. An important limitation of this study, as the authors themselves acknowledged, was the short time period assessed, which makes it difficult to draw any definitive conclusions. The short study period also made it impossible to consider the cumulative effect on patient care of fatigue over weeks and months. Concerns about health care professionals' attitudes toward errors, teamwork, and the effects of stress and fatigue on performance are prime areas for further research and improvements in health care (Sexton et al., 2000).

## 3.2 The cognitive influence on error-generating behavior

Apart from system factors, another factor leading to erroneous decisions is impaired human cognition. To a large extent, it may be the primary area for research and improvements. The study of medical error may have greater association with cognitive sciences rather than medical sciences. A majority of health care errors are due to human action or inaction caused by cognitive failure. Operating at the sharp end of the pyramid (see ▶Fig. 3.1), health care professionals are influenced to a maximum extent by these cognitive factors, which can affect anyone (Reason, 1990). As Reason (1990) notes: "Error is not the monopoly of an unfortunate few." Human performance has been classified into three categories: skill-based, rule-based, and knowledge-based (Rasmussen and Jensen, 1974). Skill-based performance includes patterns of thoughts and actions that are governed by previously stored patterns of preprogrammed instructions and those performed unconsciously. Rule-based performance includes solutions to familiar problems that are governed by rules and preconditions. Knowledge-based performance is used when new situations are encountered that require conscious analytic processing based on stored knowledge. Reason (1990) has classified skill-based errors as "slips," which are defined as unconscious aberrations in a normal routine activity. Rule-based and knowledge-based errors are termed "mistakes." Cognitive skills are closely knit in medical decision making. The technological advances in medicine and research have outpaced developments in cognitive decision making and have not been complementary. Kassierer (1995) highlighted this concern and emphasized the importance of developing these critical cognitive skills in medical students through imaginative teaching. The lack of special training for clinicians in cognitive theories has been lamented (Croskerry, 2000b). Understandably, training in cognition is not that easy. It is argued, however, that our existing knowledge of these cognitive processes is minimal, and better understanding of them may help us learn about cognitive errors and to devise teaching strategies to help avoid them (Croskerry, 2000a). Critical thinking is not part of the traditional medical school curriculum; it is assumed that medical students know how to make decisions. Recently, however, innovative training strategies in cognitive skills development through strategic management simulation techniques have been proposed (Satish and Streufert, 2002).

Cognitive decision-making errors are encountered in complex setups like health care because of biases. Leape (1994) suggested that in clinical situations that do not follow a normal pattern and that require rule-based or knowledge-based solutions, humans are biased to search for a "prepackaged solution," that is, a rule, before resorting to more strenuous "knowledge-based functioning." Bornstein and Emler (2001) noted the unavailability of any perfect method for eliminating biases in clinical decision making but suggested employing evidence-based medicine approaches and incorporation of formal decision analytical tools to improve the quality of the clinical reasoning. There can also exist a mindset that is based on a misunderstanding of the situation. There can be a disconnect between what someone believes a situation to be and what the situation really is or how it has developed (Dekker, 2002). Similarly, people can continue with their original plan despite indications that the plan should be changed. Use of modern information technology tools to improve quality and patient safety to aid in complex medical decision-making may be a feasible approach (Bates and Gawande, 2003; Kushniruk, 2001).

## 3.3 Conclusion

The decline in quality of care and the increase in the number health care errors require appropriate action. The solutions may be both sophisticated and simple. Accepting human fallibility in health care systems can be a sound beginning that allows us to address the need for safer system designs and human factor re-engineering. Health care organizations should urgently implement the best-known practices in patient safety and error reduction and simultaneously focus research strategies to attain the next level of knowledge in error prevention techniques. At the same time, adherence to basics and interdisciplinary experience, sharing, and learning should be encouraged. Health care needs to adopt systematic approaches to absorb and eliminate errors. Technical competence is not the key, but system re-engineering to enhance quality care should be the prime objective. If errors cannot be reduced to zero, the system must at the least aim to reduce to zero the number of instances when error harms a patient (Nolan, 2000). It is our hope that the study of health care errors will indeed be the discipline through which health care will advance and achieve its quality goals.

## References

Bates DW, Gawande AA. Improving safety with information technology. N Engl J Med 2003;348:2526–2534.

Blumenthal D. Making medical errors into "medical treasures." JAMA 1994;272:1867–1868.

Bornstein BH, Emler AC. Rationality in medical decision making: a review of the literature on doctors' decision-making biases. J Eval Clin Pract 2001;7:97–107.

Combs A, Taylor C. The effect of the perception of mild degrees of threat on performance. J Abnorm Soc Psychol 1952;47:420–424.

Cook RI, Woods DD. Operating at the sharp end: the complexity of human error. In: Bogner MS, ed. Human Error in Medicine. Hillsdale, NJ: Erlbaum; 1994:255–310.

Croskerry P. The cognitive imperative: thinking about how we think. Acad Emerg Med 2000a;7:1223–1231.

Croskerry P. The feedback sanction. Acad Emerg Med 2000b;7:1232–1238.

Davydov L, Caliendo G, Mehl B, Smith LG. Investigation of correlation between house-staff work hours and prescribing errors. Am J Health Syst Pharm 2004;61:1130–1134.

Dekker, S. The Field Guide to Human Error Investigations. Aldershot: Ashgate; 2002.

Dinges DF, Pack F, Williams K, et al. Cumulative sleepiness, mood disturbance, and psycho-motor vigilance performance decrements during a week of sleep restricted to 4–5 hours per night. Sleep 1997;20:267.

Easterbrook JA. The effect of emotion on cue utilization and the organization of behavior. Psychological Rev 1959;66:183–201.

Fischer JE. Continuity of care: a casualty of the 80-hour work week. Acad Med 2004;79: 381–383.

Gaba DM, Howard SK, Jump B. Production pressure in the work environment. California anesthesiologists' attitudes and experiences. Anesthesiology 1994;81:488–500.

Helmreich RL. On error management: lessons from aviation. BMJ 2000;320:781–785.

Hopkins GE. The Airline Pilots: A Study in Elite Unionization. Cambridge, MA: Harvard Univ. Press; 1971:116–141.

Institute of Medicine. Crossing the Quality Chasm: A New Health System for the 21st Century. Washington, DC: National Academy Press; 2001.

Kassirer JP. Teaching problem-solving – how are we doing? N Engl J Med 1995;332:1507–1509.

Kushniruk AW. Analysis of complex decision-making processes in health care: cognitive ap-proaches to health informatics. J Biomed Inform 2001;34:365–376.

Leape LL. Error in medicine. JAMA 1994;272:1851–1857.

Nolan TW. System changes to improve patient safety. BMJ 2000;320:771–773.

Office of Technology Assessment. Biological Rhythms: Implications for the Worker. Washing-ton, DC: Government Printing Office; 1991. Report No. OTA-BA-463.

Rasmussen J, Jensen A. Mental procedures in real life tasks: a case study of electronic trouble shooting. Ergonomics 1974;17:293–307.

Rasmussen J, Pedersen OM. Human Factors in Probabilistic Risk Analysis and Risk Manage-ment. Operational Safety of Nuclear Power Plants, vol. 1. Vienna: International Atomic Energy Agency; 1984.

Reason J. Human Error. Cambridge: Cambridge University Press; 1990.

Roberts JS, Coale JG, Redman RR. A history of the joint commission on accreditation of hospitals. JAMA 1987;258:936–940.

Romanchuk K. The effect of limiting residents' work hours on their surgical training: a Cana-dian perspective. Acad Med 2004;79:384–385.

Sandal GM. The effects of personality and interpersonal relations on crew performance during space simulation studies. Life Support Biosph Sci 1998;5:461–470.

Satish U, Streufert S. Value of a cognitive simulation in medicine: towards optimizing decision making performance of healthcare personnel. Qual Saf Health Care 2002;11:163–167.

Schwartz RJ, Dubrow TJ, Rosso RF, Williams RA, Butler JA, Wilson SE. Guidelines for surgi-cal residents' working hour. Intent vs. reality. Arch Surg 1992;127:778–782 [discussion 782–783].

Sexton JB, Thomas EJ, Helmreich RL. Error, stress, and teamwork in medicine and aviation: cross sectional surveys. BMJ 2000;320:745–749.

Shine KI. Health care quality and how to achieve it. Acad Med 2002;77:91–99.

Whang EE, Mello MM, Ashley SW, Zinner MJ. Implementing resident work hour limitations: lessons from the New York State experience. Ann Surg 2003;237:449–455.

# 4 Medical error and strategies for working solutions in clinical diagnostic laboratories and other health care areas

The IOM report (2001) stated that the prevalence of medical errors is high in today's health care system. Some specialties in health care are more risky than others. A varying blunder/error rate of 0.1–9.3% in clinical diagnostic laboratories has been reported in the literature. Many of these errors occur in the preanalytical and postanalytical phases of testing. It has been suggested that the number of errors that occur in clinical diagnostic laboratories is smaller than the number of errors that occur elsewhere in a hospital setting. However, given the volume of laboratory tests used in health care, even this low rate may reflect a large number of errors. The surgical specialties, emergency rooms, and intensive care units have been previously identified as areas of risk for patient safety. Though the nature of work in these specialties and their interdependence on clinical diagnostic laboratories present abundant opportunities for error-generating behavior, many of these errors may be preventable. Appropriate attention to system factors involved in these errors and the use of intelligent system approaches may help control and eliminate many of these errors in health care.

## 4.1 Introduction

Patient safety is a current public and health care concern. It has received substantial attention following the release of two recent reports from the IOM (2001; Kohn et al., 2000). Both of these reports have placed special emphasis on enhancing patient safety in today's health care system. Much of the problem in the provision of better and safer health care is a result of preventable adverse events due to medical errors (Kohn et al., 2000). We have previously highlighted some of the key concepts in medical error generation and have suggested a systems approach in medical error reduction (Kalra, 2004).

Some specialties in health care are more prone to errors than others. The notability of surgical specialties with respect to the rate of adverse outcomes and errors in surgical practice has been outlined earlier (Leape et al., 1991). The Quality of Australian Health Care Study (QAHCS) reported that more than half of all the adverse events recorded in the study were associated with a surgical operation (Wilson et al., 1995). The Harvard Medical Practice (HMP) Study II identified certain sites of health care that are high-risk zones for patient safety. In their analysis, the authors suggested that among these sites are operating rooms, in-patient rooms, emergency rooms, and intensive care units. The authors did not clearly delineate the adverse events occurring in the operating rooms, and presumably these may be a result of either anesthetic or surgical interventions.

However, the authors noted the low preventability of such incidents compared to events occurring in other high-risk zones, such as emergency departments (EDs) and intensive care units (ICUs), where both preventability and the potential for disability resulting from the adverse events are high. It is these high-risk zones, where preventability of adverse events is high, that are prime areas of concern and that we will discuss in greater detail. Apart from the specialties of emergency and intensive care medicine, which are highly interdependent on clinical laboratories, we also review the impact of errors in clinical diagnostic laboratories.

## 4.2 Clinical diagnostic laboratories

Laboratory reporting has a great influence on clinical decision making. With this high degree of influence, the quality of laboratory testing and reporting is of utmost importance. The IOM report (2000) has far-reaching implications for all disciplines, including pathology activities and laboratory medicine. It is to the credit of the laboratory medicine specialty that the error rates in laboratory activities are far lower than that seen in overall clinical health care (Leape, 2002). This quality, however, does not match up to industrial quality standards (Nevalainen et al., 2000). With clinicians' decision making overwhelmingly dependent on laboratory reporting, and often based solely on these reports, it becomes essential for laboratory medicine to set high quality standards and to serve as an example for other specialties to follow.

Laboratory medicine is unique in practice, and this is particularly true of clinical chemistry laboratories. Unlike many other medical processes, activities in laboratory medicine are precisely defined and therefore are more controllable than procedures or treatments in an emergency department. Laboratory medicine enjoys another unique advantage in that the specialty has pioneered statistical quality control (QC) activities that are currently used. This specialty is leagues ahead of other clinical specialties in employing quality improvement initiatives. In spite of this advantageous position, there is concern about the high number of errors in clinical diagnostic laboratories that is reported in the literature.

Statistical quality control was first introduced in clinical laboratories by Levey and Jennings in 1950. Laboratory medicine has since used statistical techniques to enhance its performance in terms of quality. The precise magnitude of the error rate in laboratory medicine has been difficult to estimate, principally because there is no definite and universally accepted definition of error. The other reasons are underreporting and impaired error detection techniques themselves. However, the error rates in laboratory medicine have been studied and documented.

McSwiney and Woodrow were one of the first to study the types and frequency of errors in a clinical laboratory in 1969. The authors reported an error rate of 2% to 3%. Chambers et al. (1986) detected a blunder rate of 0.3% in a large biochemistry laboratory. A major limitation identified in this study was the exclusion of errors occurring in the preanalytical phase of testing. Boone (1990) derived his data from hematology testing and suggested that one error occurred for every 1,000 laboratory events. Kazmierczak and Catrou (1993) reported an approximate error rate of 9.36% in their study of 438 results of replicate creatinine analysis. The small sample size and the analysis of only one analyte were the primary limitations of this study. Lapworth and Teal (1994) reported a blunder rate of 0.1% for laboratory requests. The low error rates in the study may have resulted from the fact that the authors examined only errors detected or reported in the final checking-out phase. The errors in the preanalytical, analytical, and final validation stages of the laboratory process were not included in the study. The authors also acknowledged that some errors might have gone unnoticed, as the authors did not make any special efforts to scrutinize reports intensely. Another study identified a crude rate of 1.1 problems per 1,000 patient visits in laboratory testing in primary care physicians' offices (Nutting et al., 1996).

Khoury et al. (1996) studied the error rates in Australian chemical pathology laboratories and reported error rates as high as 39% for transcription and 26% for analytical results, with the best laboratory performing error-free business only 95% of the time. Khoury et al. surveyed clinical pathology laboratories of five Australian states to compile their data. A transcription error rate as high as 39% of all requests has the potential to seriously compromise patient identification data. The authors utilized the National Quality Award Criteria for Benchmark Indicators to identify errors in the laboratory process (Australian Quality Council, 1994). Lord (1990) underlined the importance of managing risk in clinical pathology laboratories by ensuring proper identification of specimens and prevention of adverse patient care-related incidents.

Recently, we studied sample labeling errors, utilizing a retrospective audit of all incidents reported at the front end of a clinical microbiology laboratory during the period from May 1996 to December 2002 (Kalra et al., 2004a). We observed that unlabeled and mislabeled samples contributed to the majority of risk problems in the study period. However, after implementation of a zero tolerance policy to address the labeling errors in 1999, there was a gradual decrease in these events from 100% (43/43) in 1998 to 64% (16/25) of the total risk problem in 2002. Along with this, there was a gradual increase in sample rejections, which decreased the overall probability of risk events. Though our study observed only one specific aspect of laboratory testing, it underlines the effect of category-specific risk management policies in addressing the risk issues of clinical laboratories. The mislabeling rate at Johns Hopkins Hospital was found to be 1.4%, or 1 in 71 specimens (Lumadue et al., 1997). A mislabeled sample is a received sample that does not meet the hospital's criteria for sample acceptance. Medical semiautomated patient identification systems have had radical success in reducing mislabeling rates to about 0.01%, or 1 in 10,000 specimens (Astion, 2010). Implementation, however, took around seven years at Valley Hospital in New Jersey.

Hofgartner and Tait (1999), in a study based on error rates in clinical genetic testing laboratories over a 10-year period, detected error rates of 0.38% in the inspected laboratories. In this study, an onsite review of two participating laboratories was carried out, along with an extensive review of problem cases that occurred during the study period. Though the study utilized laboratory data from a wide range of medical specialties, the study design had its own limitations. The error detection strategies were far from perfect, and, as a result, some analytical errors may have been ignored if the test results were consistent with other available clinical data. It should also be emphasized that the reviewed studies had different study designs and were therefore incomparable in many aspects. Also, with no universal definition of laboratory error/blunder and imperfect error detection methods, not all laboratory errors may have been accounted for in these studies.

Bonini et al. (2002) studied and reported on the contrasting error rates in inpatients and outpatients subjected to a laboratory test. The authors suggested an error rate of

0.60% and 0.039%, respectively, in the two categories and attributed the large difference to human factors related to skill in drawing blood and the sheer amount of laboratory usage for inpatients. Another important factor contributing to erroneous laboratory reporting, which may compromise quality care and hamper patient safety, is analytical immunointerference in laboratory assays. This problem has been highlighted in two recent reports and can lead to unnecessary consultations and other interventions to patients (Ismail et al., 2002; Marks, 2002). The importance of information and adequate communication between the clinician and laboratory technologists in such instances cannot be overemphasized.

Goldschmidt and Lent (1993) observed that 12.5% of laboratory errors lead to an erroneous medical decision, 75% of the results were within normal reference limits, and in the remaining 12.5% the results were too absurd to be considered for clinical decisions. Plebani and Carraro (1997) suggested that 74% of the laboratory mistakes did not influence patients' outcomes and 19% resulted in increased costs to the patient because they led to further inappropriate investigations. The authors suggested that 6.4% of laboratory mistakes caused inappropriate care or inappropriate modification of therapy to the patient. ▶ Fig. 4.1 summarizes the frequency of errors observed in clinical laboratories (Boone, 1990; Chambers et al., 1986; Lapworth and Teal, 1994; Nutting et al., 1996; Plebani and Carraro, 1997).

## 4.3 Errors in different stages of analysis

A standard laboratory process is usually divided into three stages: preanalytical, analytical, and postanalytical. The quality cycle in a laboratory is dependent on the control of the three stages. In addition to the analytical stage, the preanalytical and postanalytical stages of a laboratory testing process also exert influence over the ultimate quality of laboratory reporting.

In the preanalytical stage, the specimen is processed before its results are analyzed. During this stage, a number of errors may occur: clinicians can forget to collect the correct specimens; specimens can be mislabeled, unlabeled, or labeled illegibly; specimens can be placed in the wrong container and lost; the specimen taken can be suboptimal or ruined (e.g., the blood sample may be clotted or hemolyzed); or the data entry may be erroneous (e.g., the wrong test may have been ordered or the wrong patient's data may have been taken).

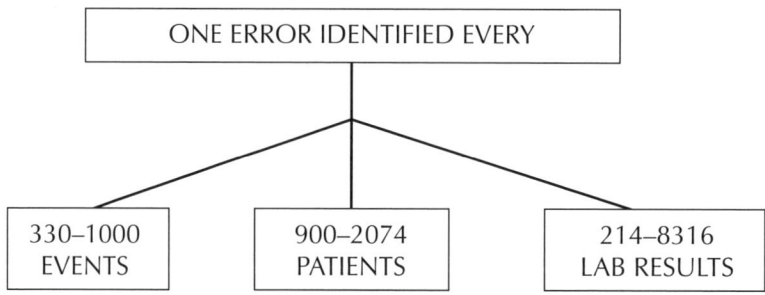

**Fig. 4.1:** Frequency of errors in clinical laboratories.

If one individual is frequently committing more errors than his peers without any obvious impairment or recklessness or his behalf, then he will have to be moved to a different service or terminated. If the errors are due to known external factors, such as bereavement, then a temporary leave from duties of patient care is necessary.

The various types and the rate of errors occurring at these three stages of a laboratory process as suggested by two studies (Plebani and Carraro, 1997; Wiwanitkit, 2001) are summarized in ▶Fig. 4.2. It has been suggested that a majority of errors occurring in a laboratory process are due to preanalytical factors (Hofgartner and Tait, 1999; Khoury et al., 1996; Plebani and Carraro, 1997; Wiwanitkit, 2001). The high rates of pre- and postanalytical errors undermine the quality performance of the analytical process and necessitate the active involvement of nonlaboratory personnel, particularly phlebotomists and clinicians, in improving the quality of laboratory reporting. The high error rates in the nonanalytical stage, where human involvement is maximal, reaffirm the susceptibility of a manual process to error.

Consider, for instance, the case of a 36-year-old female patient who is experiencing her first pregnancy. A blood test shows that she has low levels of protein-S, and her physician warns her about the significant risk of thrombosis – a rare blood disease in which blood clots form within a vein, causing significant risk to the patient. Fearing for her safety, the woman terminates the pregnancy. In this case, while the preanalytical and analytical phases have proceeded without error – the blood sample was properly prepared and tested – the physician is unaware that low protein-S values in most pregnancies do not increase the risk of thrombosis. This is a postanalytical error, as the correct results have been incorrectly analyzed.

Such mistakes are costly. Patients suffer harmful physical and emotional consequences from receiving misdiagnoses and undergoing unnecessary or even dangerous procedures. This, in turn, increases the cost of care in the form of additional lab tests

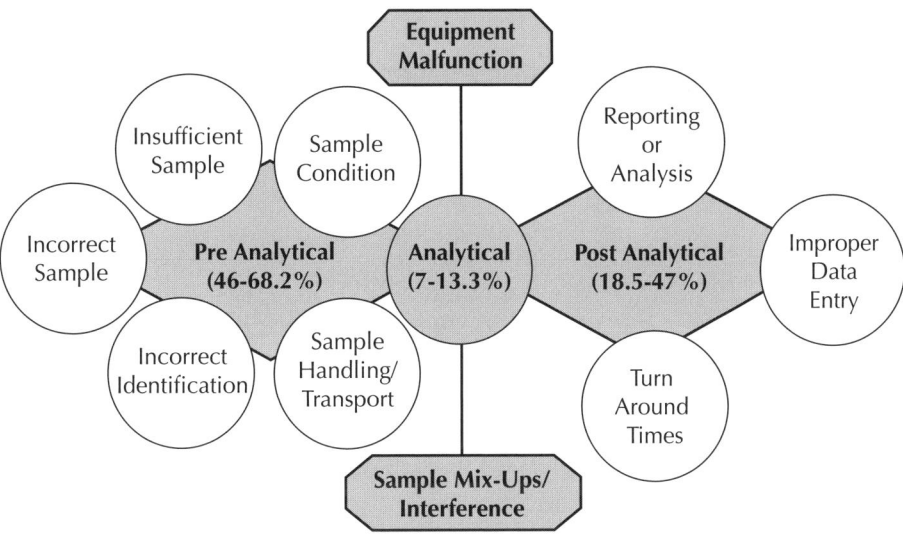

**Fig. 4.2:** Types and rates of error in the three stages of the laboratory testing process.

and professional labor to rectify the resulting medical errors, as well as costing valuable physician and laboratory staff time in the corrective action taken after incorrectly interpreted tests, as well as the interpretation itself.

Laboratory medicine has high quality standards. This is ensured by a number of regulatory authorities, including the National Joint Commission on Accreditation of Health Care Organizations (JCAHO)) and the College of American Pathologists (CAP), that periodically outline the quality and accrediting standards required for practice. In addition, quality improvement initiatives, such as the Q-probes quality improvement program of the CAP (Bachner and Howanitz, 1991), instituted to establish provisional benchmarks for measuring systematic quality improvement efforts, are also in place. The efforts of such initiatives are paying high dividends in areas that have been identified for quality improvement and enhancing patient safety, such as wristband patient identification and laboratory specimen acceptability (Howanitz et al., 2002; Zarbo et al., 2002). Other professional societies are also involved in sponsoring a number of educational programs that focus on quality issues. The professional boards responsible for certification processes have ensured the competency in practice of quality laboratory medicine over the years. These measures have helped laboratory medicine to function at the level of reliability that is expected from any clinical practice involved in health care. However, the focus of these certification and accreditation processes is most often on laboratories' general performance, rather than particularly on errors (Bonini et al., 2002) and other related issues closely linked to quality.

Two particularly important quality issues confronting modern laboratory services are the problems of weak control over affairs in the preanalytical stage and the identification of errors and subsequent reporting. The preanalytical phase is clearly the most important part of the laboratory testing process. What a laboratory analyzes and ultimately reports is a direct result of the quality of the specimen or sample delivered for testing. However, it is suggested that the clinical laboratories assume responsibility for the whole cycle of the testing process, from the physician's ordering a laboratory investigation to the recognition of the significance of the reported result in the management of the patient. This responsibility requires that the testing cycle be completed by involving other professionals in the quality loop. The benefits achieved by such efforts have previously been demonstrated. Renner et al. (1993) reported the influence of phlebotomists on improving quality and patient safety in transfusion practice by monitoring the level of conformity of wristbands to quality standards.

## 4.4 Strategies for identification and prevention of errors

The other issue facing today's laboratory medicine practice is the identification of errors itself. This issue has previously been highlighted in the contexts of clinical laboratories and anatomic pathology. Bonini et al. (2002) highlighted this peculiar problem for laboratories and emphasized its difficulty, because errors may not necessarily produce detectable abnormal results or raise questions in the ultimate user of the laboratory results. Similar concerns have been highlighted in anatomical pathology because of the difficulty in addressing what exactly constitutes a diagnostic error. Sirota (2000) suggested that the failure of two independent pathologists to reach the same diagnosis in a specific case might not represent a diagnostic error in itself and might be a simple case of interobserver variation. However, such variations can have far-reaching consequences for a patient because of the different diagnoses and subsequent case management. There are some other common uncertainties in a laboratory testing process over which the clinical laboratories have limited control. These may be in the form of biological variation, systematic errors, analytical variation, and interfering substances that are not controlled even by the addition of blocking agents.

Many of the quality improvement and patient safety measures used in different sectors of health care may also apply widely to the practice of laboratory medicine. We have previously suggested a "no-fault" error reporting model for pathology and laboratory medicine, which emphasizes the importance of risk-free voluntary error reporting systems for health care practitioners to develop a safe health care system (Kalra et al., 2003). Further elaborating on this model, we have also integrated an incident reporting system in our clinical chemistry laboratory. The incident reporting system includes various types of errors in the clinical laboratory and encourages voluntary reporting of critical laboratory incidents on a nonpunitive basis. The implemented model also includes an educational program and a quality care committee to monitor the critical incidents and medical errors in the clinical laboratory (Kalra et al., 2004b). The quality care committee reviews the reported errors periodically and recommends appropriate action aimed at preventing the occurrence of similar events. This model, along with the laboratory quality assurance program, forms an integral part of clinical diagnostic/chemistry laboratories. The model is in its infancy, and its impact is being continuously assessed.

Certain other measures with particular relevance to laboratory medicine initiatives in improving quality care are reviewed in this section. Apart from making patient safety a priority and a departmental goal, laboratory medicine should dedicate resources for education and research to address the unique concerns in practices like those discussed in the preceding paragraph. Research initiatives should be directed toward establishing the best evidence-based practices and diagnostic gold standards. Studies should also be directed toward eliminating biases in decision making related to diagnostic activities. Clinical laboratories should identify areas where human involvement in the process can be minimized and increased possibilities for automation and robotics exist. Laboratory automation provides for standardized workflow and helps eliminate many error-prone steps undertaken by humans. Automation may decrease opportunity for errors caused by active human factors like stress, fatigue, negligence, and cognitive impairment.

Health care is an intensive and highly skilled profession, and clinicians are required to work long hours. Moreover, the work situation itself is particularly prone to fatigue-associated error. For example, clinicians are required to detect rare but important signals; for example, they must notice error flags in a system of automated tests. This is in addition to performing tasks that require multitasking and creative troubleshooting, all of which are susceptible to fatigue (Astion, 2009).

There are, however, both temporary and long-term ways to reduce fatigue. Stimulants, such as coffee; brighter lights in work areas; and conversation and physical activity among staff are all ways to temporarily reduce fatigue (Astion, 2009). In the long term, staggered shifts should be implemented, rather than abrupt shift changes. The workers who have just begun their shift should be made to handle the hardest work available. Staggered shifts allow for better hand-offs of workloads than abrupt shifts. When clinicians are transferring their workload at the end of their shift, they're not only tired but also eager to depart from the workplace. As such, the quality of information received by the incoming clinician isn't as high as it would be were the departing colleague less tired, and communication breakdowns would be less common. Furthermore, there should exist a permanent night shift so that workers can develop regular sleep cycles.

The emergency departments and intensive care units depend heavily on clinical laboratories. This dependence involves but is not restricted to therapeutic drug monitoring, interpretation of laboratory results in view of interfering substances, and laboratory monitoring or surveillance to predict signs of clinically unrecognized toxicity. The other critical area where the clinical laboratories and EDs are integrated to achieve effective clinical decision making is diagnosis. This issue, in the context of acute myocardial infarction (AMI), is discussed in the following section. With greater integration and effective information sharing between clinical laboratories and these units, the medication and diagnostic errors occurring in the EDs and intensive care units may be reduced to a large extent.

## 4.5 Errors in emergency medicine

The chaos, the trauma, and the crises witnessed in the emergency departments (EDs) are like no other in health care. EDs may be seen as the diametric opposite of the highly controlled environment of the clinical diagnostic laboratory setting. The ED environment provides abundant opportunities for generating medical errors. The HMP Study II claims that nearly 3% of all adverse events occur in the EDs. More important, it has been suggested that emergency medicine is a crucial area for preventable medical errors (Chin et al., 1999). The IOM report (Kohn et al., 2000) suggests that the highest rate of medical errors occurs in EDs. This assumes significance when one considers that another study identifies EDs as having the highest rate of negligent adverse events, and approximately 95% of adverse events attributed to emergency physicians were judged negligent (Thomas et al., 2000). Similar findings have been reported earlier (Leape et al., 1991).

"Quality emergency care is a fundamental right and should be available to all those who seek it" states the American College of Emergency Physicians' value statement. Through this statement, the college has prioritized quality care and emphasizes that it will accept no compromise in patient safety, whatever the pressures faced by the ED staff. Safety, timely care, and quality care form the three main goals of an emergency care setting. In all fairness, every ED professional strives to achieve the best possible health care results for their patients, but some deficiencies may nonetheless exist.

There is a fundamental difference between the error-influencing environment of EDs and the atmosphere in other medical wards: the time constraint under which the ED professional works. Other contributing factors may include insufficient accessibility to medical history, varying rates of patient inflow into EDs, and the fact that EDs encounter more high-risk patients who require more procedural interventions, thereby presenting a greater risks of errors, than do other services. However, the emergency medicine system is expected to thrive under such demanding conditions. Given these constraints, efforts to take EDs toward quality improvements in care must come through adequate staffing, use of modern information technology, and specialized training initiatives for error reduction through education and professional development programs.

The reliability of the figures quoted by studies to suggest the error rates in EDs can itself be questioned. Emergency medicine is a relatively new specialty, and the figures quoted by the HMP study II and the IOM report are derived from fairly old data, data that were gathered before trained and board-certified ED physicians were managing and providing care in EDs. Times have changed. ED personnel are now professionally staffed and managed, but one aspect of EDs remains the same – the time-pressured environment. This environment is a fertile breeding ground for errors. Hence, it may not be appropriate to assume that the statistics on error rates in EDs are exaggerated and that EDs are safe. The reality is otherwise.

Emergency medicine was one of the first specialties to realize the potential of errors to compromise patient safety and quality of care. The emergency medicine specialty was the pioneer in patient safety initiatives. The wide varieties of presentations and diagnoses in EDs have led to focused research and effective interventions. Rosen et al. (1983) were among the first researchers to study errors in emergency medicine, and they suggested innovative approaches to help eliminate such errors. Karcz et al. (1996) reviewed malpractice claims filed against emergency physicians in Massachusetts over an 18-year period, 1975–1993. The authors identified chest pain, abdominal pain, and fractures as areas where physicians were at high risk of making errors.

Chest pain as a result of acute cardiac ischemia is both a common and a frequently missed diagnosis in EDs. Approximately 2% of patients with AMI were misdiagnosed, with 25% of these patients having a fatal outcome, in a multicenter study in a coronary care unit (McCarthy et al., 1993). The authors employed a case control study design with a sample size of 1,050 patients with AMI. They compared patients with a missed AMI diagnosis to two control groups, patients admitted with AMI and a group discharged without AMI. At least 50% of misdiagnoses could have been prevented if the ST elevations of electrocardiographs (ECGs) had been recognized or patients who were recognized as having ischemic heart disease by the physicians in the ED had been admitted. However, the authors suggested that the patients with missed AMI were significantly less likely to have ECG changes than patients admitted with AMI. Chan et al. (1998), by means of a retrospective analysis, estimated that at least 27% of AMI were missed in the EDs and cited the absence of chest pain and the lack of ST elevations in ECGs as the main predisposing factors for misdiagnosis. The relative higher rate of missed AMI diagnoses in this study than in the earlier study by McCarthy et al. (1993) may be attributed to a smaller sample size (159 patients) in the Chan et al. study and the use of a different study design. Another reason for the difference may be the demographic and regional variations evident in the two study designs. The former was conducted in North America and the latter in Asia. The preventability rate in the study by Chan et al. was 35%, and the errors were mostly attributed to inaccurate interpretation of ECGs. However, the competence of emergency physicians in ECG interpretation of patients with chest pain and ST elevation has been proved (Brady et al., 2000). Brady et al. also suggested that the low rate of ECG misinterpretation was of minimal clinical consequence.

In 2005, the Canadian Patient Safety Institute (CPSI) launched a grassroots campaign named *Safer Healthcare Now!* (SHN). The campaign aims to promote patient safety by endorsing six targeted, evidence-based interventions by health care professionals (CPSI, 2007). One of the promoted interventions is improved care for AMI in Canada. CPSI aims to help hospitalized patients by ensuring the reliable delivery of evidence-based care, including a focus on correct diagnosis of the illness. A National Steering Committee officially leads the SHN campaign, with support from the CPSI.

Pope et al. (2000) found a correlation between racial, gender, and clinical factors and failure to hospitalize patients presenting with acute cardiac ischemia. The authors suggested that nonwhites, women less than 55 years old, patients presenting chiefly with shortness of breath, and patients with nondiagnostic ECGs were ideal candidates for nonhospitalization. All of these findings point to other extraneous sources apart from human factors in erroneous decision making. These factors are more important as they are unidentified and hence have the potential to recur and to lead to more catastrophic consequences in the future. Clinical laboratories have a limited but specific role in minimizing these missed diagnoses. The judicious use of new-generation cardiac troponin tests based on evidence-based medicine may be a solution. In establishing the cut-off levels for therapeutic interventions, the clinical laboratories have to play a crucial role along with EDs and cardiology units. Clear communication and interaction between clinical laboratories and clinicians are essential in maximizing the availability and effectiveness of these and other tests in helping them to make better decisions. Clinical laboratories should also focus on decreasing the turnaround times (TAT) for diagnostically important tests like those used for AMI. The effective use of clinical laboratories, together with ECGs and other useful risk factor information, may serve to enhance clinical decision making and eliminate errors in the diagnosis of AMI.

Another subgroup of diagnoses that is often misdiagnosed in EDs is acute surgical emergencies, particularly appendicitis. This condition often poses problems for accurate diagnosis and prompt treatment because of the differing clinical features in every patient. Rothrock et al. (1991) compared the clinical presentations of children with misdiagnosed appendicitis and those in whom the appendicitis was correctly diagnosed the first time. These investigators concluded that clinical presentations in the two groups were different. Reynolds (1993) retrospectively reviewed hospital ED records and reported that at least 7% of the patients were seen twice in their EDs before the diagnosis of appendicitis was finally made. Similar results were obtained in a retrospective case series study on nonpregnant women of childbearing age, which showed that 33% of the women with appendicitis were initially misdiagnosed (Rothrock et al., 1995). One of the main reasons for the misdiagnoses in the two studies was the absence of the typical findings that are normally associated with appendicitis. The consequences of such a delay in diagnosis can be grave. The most common complication of delayed diagnosis or misdiagnosed appendicitis is perforation. The rate of perforation can vary between 71% (Cappendijk and Hazebroek, 2000) and 91% (Rusnak et al., 1994). Other complications include the need for more extensive surgical procedures, more postoperative complications, and longer hospital stays. The problem lies not only in underdiagnosis but also in overdiagnosis of appendicitis in EDs, which results in negative appendectomies, surgeries performed as a result of presumed appendicitis. The frequency with which misdiagnosis leads to unnecessary appendectomies has not changed very much in spite of modern imaging techniques like CT scans (Flum et al., 2001). An estimated $741.5 million in total hospital costs was attributed to this presumed diagnosis of appendicitis in 1997 (Flum and Koepsell, 2002).

Another area of concern in EDs that needs further improvements in quality to benefit patient safety is radiograph interpretation. Emergency medicine professionals are often required to interpret radiographs in the absence of trained radiologists. About

3% of clinically significant errors made in an emergency medicine residency program involved the interpretation of radiographs (Gratton et al., 1990). Brunswick et al. (1996) reported that there was only 1% discordance between the readings of emergency physicians and those of trained radiologists, though 46% of these discordant misreads were deemed clinically significant. Recently, it was suggested that emergency physicians frequently missed specific radiographic abnormalities and that there was a high degree of discrepancy between their interpretations and those of trained radiologists (Gatt et al., 2003). The study noted, however, that these misreadings had relatively few clinical implications. It is generally felt that emergency physicians perform fairly well at interpreting plain radiographs but occasionally miss significant findings, leading to adverse outcomes. Quality improvement initiatives in reading radiographs may include the presence of a trained radiologist for consultation around the clock, but, because of the small number of significant misinterpretations, this may not prove to be cost effective (Klein et al., 1999). Nevertheless, a higher diagnostic efficiency is established when the emergency physician's interpretation of a chest radiograph is complemented by that of a radiologist (Urrutia et al., 2001). Shirm et al. (1995) have suggested that a quality-assurance system that recalls patients who may require a change in treatment appears to be an effective method of patient management only when discordant interpretations are identified and promptly acted upon. Eng et al. (2000) documented that high-quality image displays and trained radiologist coverage for EDs can improve the accuracy of interpreting ED radiographs. Mann and Danz (1993) suggested that residents posted exclusive to nighttime rotations performed higher quality interpretations of radiographs than did residents who were covering EDs on call after a full day's work. This particular finding adds credence to the argument that the resident work-hour limitations that have been recently imposed by the ACGME enhance overall patient care (Romanchuk, 2004).

## 4.6 Errors in intensive care medicine

A majority of general hospitals have fully functioning intensive care units (ICUs). These units are managed by professionals from multiple disciplines and provide care for critically ill patients with life-threatening diseases or injuries. Provision of timely, high-quality care significantly enhances the chances of survival for these critically ill patients. Though ICUs may be further subdivided to cater to a specific subgroup of patients (e.g., surgical ICUs [SICU], pediatric ICUs [PICU], and neonatal ICUs [NICU]), the basic work profile remains the same, and so does the potential for error. It is for this reason all the subunits of ICUs are collectively referred to as ICUs in this review.

Modern intensive care units are associated with complex care provided to critically ill patients. The complexity of care provided is compounded by understaffing, leading to increased workloads in ICUs. This setting may provide an ideal environment for errors to happen, which can have serious consequences for those receiving care.

So how safe are present-day ICUs? If the reviewed literature is to be believed, then the answer is: not very safe. Abramson et al. (1980) were among the first to study the rates of adverse events in intensive care units. The authors retrospectively analyzed adverse incident reports over a five-year period. The authors found a total of 145 adverse incidents filed in a general ICU. The study reported that the mortality of patients with an incident report filed during their ICU admission was 41%, whereas the rate for all ICU patients was 21%. A prospective two-center study of adult ICUs reported iatrogenic complications in 31% of the total of 400 admissions, and 13% of these admissions had major complications as a result of the adverse event (Giraud et al., 1993). The study also reported that the mortality risk was close to twice as high in patients with a major iatrogenic complication as in other patients without any iatrogenic complications. An important limitation of this study was that it excluded medication-related adverse events. Exclusion of this important category of adverse events, as discussed later, may have led the researchers to underestimate the adverse event rates in the study. Donchin et al. (2003) investigated the nature and causes of human errors in ICUs through a concurrent-incident study. The study suggested that an estimated 1.7 errors occurred per patient per day, and for the whole ICU it reported that a severe or potentially detrimental error occurred, on average, twice a day. The authors cautioned about generalizing from these results as the study adopted a direct-observation methodology, which may have prevented some of the error-generating behavior of the staff. Another limitation of the study was that it was conducted in a unit that was grossly understaffed and that also served as a training facility. These factors may have influenced the results obtained by the study.

A high proportion of errors occurring in ICUs may be associated with medications (Flaatten and Hevroy, 1999). An earlier prospective study in a NICU reported a total of 313 medication error incidents in 23,307 patient days, and these errors were responsible for a majority of all serious incidents (Vincer et al., 1989). However, negligence related to medication schedules and failure to regulate intravenous infusions were responsible for the majority of these events. Another interesting study compared the rates of preventable adverse drug events in ICUs with general care units (Cullen et al., 1997). The study revealed that the rate of preventable adverse drug events was twice as high in ICUs as in non-ICUs. The investigators conducted structured interviews that indicated that there were almost no differences between ICUs and general care units on many characteristics of the patient and caregiving teams. The authors also failed to notice any correlation between adverse drug events in the units and the working environment of the staff. An error rate of 6.6% for all drug administration events has been reported in a prospective pharmacist-performed observation study (Tissot et al., 1999). Dosing errors accounted for the majority of detected errors, and, though there were no fatal errors, at least 20% of the errors were potentially life threatening. However, the authors of the study could not correlate the high incidence of preventable adverse drug events with increased workload or a stressful environment. Nevertheless, the authors identified deficiencies in the overall organization of the hospital medication track, in patient follow-up, and in staff training as reasons for the errors. The study, like some of the others cited, suffered the limitation of observer bias that is inherent in some directly observed studies (Donchin et al., 1995).

Another observational study on medication errors of adult patients in ICUs found more promising results than those reported in earlier studies. The study reported a rather low 3.3% incidence rate for medication errors (Calabrese et al., 2001). The authors identified wrong infusion rates as the most common type of error. However, it must be mentioned that the different methodologies of various authors, their varying definitions of errors, and the different settings make fair comparison of results difficult, if not impossible. Allan and Barker (1990) suggest that the "disguised observation technique" used by some authors (Calabrese et al., 2001; Tissot et al., 1999) may be a more efficient method for medication error research than the retrospective incident report analysis used by other authors (Abramson et al., 1980). Some of the reviewed studies suffer from limitations in the form of study design, as their focus was restricted to drug administration errors (Tissot et al., 1999; Calabrese et al., 2001) or nursing staff errors (Tissot et al., 1999). The specific focus on certain events or groups may have led researchers to miss other critical incidents in different settings. Also, the error detection techniques utilized by the authors themselves may not be perfect, allowing many medication errors to escape scrutiny. Though the error rates reported in the literature vary, the consequences remain the same – patient harm.

Some of the common factors promoting errors have been identified as a lack of standardization, insufficient labeling of medications, improper documentation, and, most important, poor communication (Gopher et al., 1989). Clear communication is a principal component in providing quality care in ICUs. Fassier (2010) demonstrated that poor communication underlies many conflicts in the ICU. Studies indicate that it is generally believed that communication practiced in ICUs is suboptimal and that ICU professionals poorly understand the ultimate goals of care for any particular day (Pronovost et al., 2003). Pronovost et al. reported that fewer than 10% of ICU residents and nurses understood the goals of care. The authors implemented a daily goals form in the ICUs and observed that, after its implementation, more than 95% of nurses and residents understood the goals of care for that day. This intervention resulted in a reduction in patients' total length of stay and significant improvements in the provision of quality care.

Human involvement in errors is very high in ICUs, and at times these errors may have catastrophic consequences. Wright et al. (1991) attributed 80% of critical events to human error. Earlier, Abramson et al. had reported that 63% of incidents were due to human error. Other studies have periodically reported similar findings (Giraud et al., 1993; Stambouly et al., 1996). Bracco et al. (2001) studied human errors prospectively in a multidisciplinary ICU and found that 31% of the critical incidents reported in the study were due to human errors. A majority of the errors in the study occurred while a clinician was executing an action, but planning errors were primarily responsible for significant complications. These findings suggest that human errors are a major cause of critical incidents, threatening quality care and patient safety in an ICU environment.

Another critical area that needs attention if we are to make ICUs safer is staff workload and training. Understaffing and a decreased staff-to-patient ratio are burning concerns not only in ICUs but also health care in general. This problem with respect to ICUs deserves special mention because of the growing number of procedural interventions and the critical condition of the patients in ICUs. This shortage of staff threatens quality and service levels in ICUs. The research in this area has shown that mortality, adjusted for all other factors, was more than twice as high in patients exposed to ICUS with the greatest workloads as those where workers had lower ICU workloads (Tarnow-Mordi, 2000). It has also been reported that patients in ICUs with fewer nurses have an increased risk of postoperative respiratory complications (Dimick et al., 2001a; Pronovost et al., 2001a). Many of the workload-associated problems could be offset to an extent by employing nurses without ICU experience who could assist the regular ICU nursing staff by providing basic nursing care, thereby relieving the ICU staff's workload while not compromising the quality of care (Binnekade et al., 2003).

The other area that needs to be addressed if we are to enhance quality of care and patient safety in ICUs is staff training. Present-day ICUs commonly follow one of the three staffing models. The most common model is the open-model ICU, where the admitting physician is in charge of the care of their patients. The next most common is the closed-model ICU, which is managed by trained critical-care specialists, or intensivists, who are responsible for the full care of the patients in ICUs. The third model is the semi-closed model, in which the admitting physician coordinates with the intensivists who manage the care of the patient. ICUs have been slow to convert from the open model to the closed model, though the benefits outweigh the costs and risks. An estimated 53,000 lives and $5.4 billion in costs could be saved annually if intensivist physician staffing were implemented in the nonrural United States (Pronovost et al., 2001b). The intensivists bring with them advanced training in critical-care medicine and technical skills, which immensely benefits the patient. The evidence suggests that ICUs staffed with intensivists are more likely to produce the desired results for patient outcomes (Dimick et al., 2001b). The organizational change from open to closed ICUs improves clinical outcomes, and it has been reported that nurses had more confidence in the clinical judgment of the physician in the closed ICU than in the open ICU (Carson et al., 1996). And it is not only physicians' expertise, training, and experience that benefits quality of care and patient safety. Morrison et al. (2001) suggested that nursing staff inexperience contributed to adverse patient outcomes in ICUs. Amid all these disturbing trends in ICUs, there seems to be promise. Much of that promise derives from the premise that continuous research and incident analysis by means of critical incident reporting system will improve patient safety.

The reporting of incidents, including both adverse events and near-mistakes caught just before they actually took place, is an essential component of improving patient safety. In intensive care, the Australian Incident Monitoring System (AIMS-ICU) has pioneered efforts in this direction (Beckmann, 1996). AIMS-ICU is a national critical error reporting system set up in 1993 to assess the impact of an anonymous, voluntary reporting system on quality of care and patient safety. It has been demonstrated that critical incident reporting is effective in revealing the latent errors present in the system, the role of human error, and the generation of these critical incidents (Buckley et al., 1997). Additionally, this reporting system is a useful way to highlight problems that have been previously undetected by the routine quality assurance programs. Frey et al. (2002) reported on the benefits of critical incident reporting systems on medication error prevention in ICUs. In a study of two methods, incident monitoring and retrospective chart review, it was found that incident monitoring offered more contextual information about the incidence of such errors and identified a larger number of preventable problems than did medical chart reviews (Beckmann et al., 2003). However, the authors of the study suggested that incident monitoring identified fewer iatrogenic infections and advised supplementing this process with selective audits and focused medical chart reviews to detect those problems not reported well by incident monitoring alone.

Patient monitoring is critical in ICU environments. The purpose of monitoring equipment is to provide timely and maximum information. Good cognitive ergonomic design of monitoring equipment should serve to reduce errors caused by human factors. Instrument-redesigning research is as important in reducing human errors as is focusing on organizational issues related to work. Innovative, targeted interventions for ICU alarms and their warnings have been previously suggested. McIntyre and Nelson (1989) suggested that human voice messages were far superior to nonverbal signals emanating from ICU monitoring devices. The investigators suggested that such warning signals might influence humans to make fewer errors. Schoenberg et al. (1999) devised a module that assists in eliminating false and insignificant alarms in the ICU setting and improves the efficacy of these devices.

The high rate of medication errors in ICUs is a primary area that needs improvement. Landis (1999) reported that putting a pharmacist on the ICU team could cut ordering errors by 66%. This area requires innovative approaches so that basic procedures such as ordering, following the orders, and drug administrations are made error-proof and any potential for erring is completely eliminated. All ICUs should adopt critical incident reporting as a routine, and these reporting systems should be incorporated in the policy of continuous quality improvements. Such a system would form a vital link in identifying deficiencies in the system. Correction of these deficiencies will lead to a reduction in future incidence of these critical incidents and will promote overall patient safety.

## 4.7 Conclusion

Efforts to reduce medical errors and enhance patient safety can be directed toward many health care processes. Clinical diagnostic laboratories, EDs, and ICUs are recognized risk zones with high potential for error generation. It is prudent that maximum efforts be directed toward preventing errors in areas of high preventability. The human factors involved in generating errors in these specialties deserve priority, and appropriate technology may play a vital role in mitigating many of these factors. Staffing shortages in some of these risk zones may be an individual factor affecting the quality of care provided. Regulatory and legislative bodies have a central role in addressing many of these patient safety issues. Adopting intelligent systems approaches to promoting efficiency and enhancing team coordination to facilitate optimal outcomes in patient care are a necessity. The design and development of adequate programs to achieve a culture of safety in all health care processes will be a continuing challenge that needs to be appropriately addressed.

## References

Abramson NS, Wald KS, Grenvik AN, Robinson D, Snyder JV. Adverse occurrences in intensive care units. JAMA 1980;244:1582–1584.

Allan EL, Barker KN. Fundamentals of medication error research. Am J Hosp Pharm 1990; 47:555–571.

Astion M. Fatigue and error: an interview with Dr. Matthew B. Weinger. Clin Lab News 2009;35:10–11.

Astion M. Lean principles and patient safety overview: principles and solutions in using LEAN to reduce errors in patient ID and specimen collection. Presentation. 2010.

Australian Quality Council. Australian Quality Awards Foundation. Assessment Criteria and Application Guidelines. Sydney: Australian Quality Council; 1994.

Bachner P, Howanitz PJ. Using Q-Probes to improve the quality of laboratory medicine: a quality improvement program of the College of American Pathologists. Qual Assur Health Care 1991;3:167–177.

Beckmann U, Bohringer C, Carless R, et al. Evaluation of two methods for quality improvement in intensive care: facilitated incident monitoring and retrospective medical chart review. Crit Care Med 2003;31:1006–1011.

Beckmann U, West LF, Groombridge GJ. The Australian incident monitoring study in intensive care: AIMS-ICU. The development and evaluation of an incident reporting system in intensive care. Anaesth Intensive Care 1996;24:314–319.

Binnekade JM, Vroom MB, de Mol BA, de Haan RJ. The quality of intensive care nursing before, during, and after the introduction of nurses without ICU-training. Heart Lung 2003;32:190–196.

Bonini P, Plebani M, Ceriotti F, Rubboli F. Errors in laboratory medicine. Clin Chem 2002; 48:691–698.

Boone DJ. Comment on "Random errors in haematology tests." Clin Lab Haematol 1990; 12(Suppl. 1):169–70.

Bracco D, Favre JB, Bissonnette B, et al. Human errors in a multidisciplinary intensive care unit: a 1-year prospective study. Intensive Care Med 2001;27:137–145.

Brady WJ, Perron A, Ullman E. Errors in emergency physician interpretation of ST-segment elevation in emergency department chest pain patients. Acad Emerg Med 2000;7: 1256–1260.

Brunswick JE, Ilkhanipour K, Seaberg DC, McGill L. Radiographic interpretation in the emergency department. Am J Emerg Med 1996;14:346–348.

Buckley TA, Short TG, Rowbottom YM, Oh TE. Critical incident reporting in the intensive care unit. Anaesthesia 1997;52:403–409.

Calabrese AD, Erstad BL, Brandl K, Barletta JF, Kane SL, Sherman DS. Medication administration errors in adult patients in the ICU. Intensive Care Med 2001;27:1592–1598.

Canadian Patient Safety Institute. Safer Healthcare Now! Program Review. July 23, 2007.

Cappendijk VC, Hazebroek FW. The impact of diagnostic delay on the course of acute appendicitis. Arch Dis Child 2000;83:64–66.

Carson SS, Stocking C, Podsadecki T, et al. Effects of organizational change in the medical intensive care unit of a teaching hospital: a comparison of "open" and "closed" formats. JAMA 1996;276:322–328.

Chambers AM, Elder J, O'Reilly DS. The blunder-rate in a clinical biochemistry service. Ann Clin Biochem 1986;23:470–473.

Chan WK, Leung KF, Lee YF, Hung CS, Kung NS, Lau FL. Undiagnosed acute myocardial infarction in the accident and emergency department: reasons and implications. Eur J Emerg Med 1998;5:219–224.

Chin MH, Wang LC, Jin L, et al. Appropriateness of medication selection for older persons in an urban academic emergency department. Acad Emerg Med 1999;6:1232–1242.

Cullen DJ, Sweitzer BJ, Bates DW, Burdick E, Edmondson A, Leape LL. Preventable adverse drug events in hospitalized patients: a comparative study of intensive care and general care units. Crit Care Med 1997;25:1289–1297.

Dimick JB, Pronovost PJ, Heitmiller RF, Lipsett PA. Intensive care unit physician staffing is associated with decreased length of stay, hospital cost, and complications after esophageal resection. Crit Care Med 2001a;29:753–758.

Dimick JB, Swoboda SM, Pronovost PJ, Lipsett PA. Effect of nurse-to-patient ratio in the intensive care unit on pulmonary complications and resource use after hepatectomy. Am J Crit Care 2001b;10:376–382.

Donchin Y, Gopher D, Olin M, et al. A look into the nature and causes of human errors in the intensive care unit. 1995. Qual Saf Health Care 2003;12:143–147.

Eng J, Mysko WK, Weller GE, et al. Interpretation of emergency department radiographs: a comparison of emergency medicine physicians with radiologists, residents with faculty, and film with digital display. AJR Am J Roentgenol 2000;175:1233–1238.

Fassier T, Azoulay E. Conficts and communication gaps in the intensive care unit.Curr Opin Crit Care 2010;16:1–12.

Flaatten H, Hevroy O. Errors in the intensive care unit (ICU). Experiences with an anonymous registration. Acta Anaesthesiol Scand 1999;43:614–617.

Flum DR, Koepsell T. The clinical and economic correlates of misdiagnosed appendicitis: nationwide analysis. Arch Surg 2002;137:799–804.

Flum DR, Morris A, Koepsell T, Dellinger EP. Has misdiagnosis of appendicitis decreased over time? A population-based analysis. JAMA 2001;286:1748–1753.

Frey B, Buettiker V, Hug MI, et al. Does critical incident reporting contribute to medication error prevention? Eur J Pediatr 2002;161:594.

Gatt ME, Spectre G, Paltiel O, Hiller N, Stalnikowicz R. Chest radiographs in the emergency department: is the radiologist really necessary? Postgrad Med J 2003;79:214–217.

Giraud T, Dhainaut JF, Vaxelaire JF, et al. Iatrogenic complications in adult intensive care units: a prospective two-center study. Crit Car Med 1993;21:40–51.

Goldschmidt HMJ, Lent RW. Gross errors and work flow analysis in the clinical laboratory. Klin Biochem Metab 1995;3:131–140.

Gopher D, Olin M, Badihi Y. The Nature and Causes of Human Errors in a Medical Intensive Care Unit. Proceedings of the 33rd Annual Meeting of the Human Factors Society. Denver: Human Factors Society; 1989:956–960.

Gratton MC, Salomone III JA, Watson WA. Clinically significant radiograph misinterpretations at an emergency medicine residency program. Ann Emerg Med 1990;19: 497–502.

Hofgartner WT, Tait JF. Frequency of problems during clinical molecular-genetic testing. Am J Clin Pathol 1999;112:14–21.

Howanitz PJ, Renner SW, Walsh MK. Continuous wristband monitoring over two years decreases identification errors: a College of American Pathologists Q-Tracks study. Arch Pathol Lab Med 2002;126:809–815.

Institute of Medicine. Crossing the Quality Chasm: A New Health System for the 21st Century. Washington, DC:National Academy Press; 2001.

Ismail AA, Walker PL, Barth JH, Lewandowski KC, Jones R, Burr WA. Wrong biochemistry results: two case reports and observational study in 5,310 patients on potentially misleading thyroid-stimulating hormone and gonadotropin immunoassay results. Clin Chem 2002;48:2023–2029.

Kalra J, Medical errors:an introduction to concepts. Clin Biochem 2004:37:1043–1051.

Kalra J, Neufeld H, Knight I, et al. Risk management at the front end of a clinical laboratory: a retrospective study [Abstract]. Clin Biochem 2004a;37:724.

Kalra J, Saxena A, Mulla A, Neufeld H, Qureshi M, Massey KL. Medical error: a clinical laboratory approach in enhancing quality care [Abstract]. Clin Biochem 2004b;37: 732–733.

Kalra J, Saxena A, Mulla A, Neufeld H, Qureshi M, Sander R. Medical error and patient safety: a model for error reduction in pathology and laboratory medicine [Abstract]. Clin Invest Med 2003;26:215.

Karcz A, Korn R, Burke MC, et al. Malpractice claims against emergency physicians in Massachusetts: 1975–1993. Am J Emerg Med 1996;14:341–345.

Kazmierczak SC, Catrou PG. Laboratory error undetectable by customary quality control/quality assurance monitors. Arch Pathol Lab Med 1993;117:714–718.

Khoury M, Burnett L, Mackay MA. Error rates in Australian chemical pathology laboratories. Med J Aust 1996;165:128–130.

Klein EJ, Koenig M, Diekema DS, Winters W. Discordant radiograph interpretation between emergency physicians and radiologists in a pediatric emergency department. Pediatr Emerg Care 1999; 15:245–248.

Kohn LT, Corrigan JM, Donaldson MS, editors. Committee on Quality of Healthcare in America, Institute of Medicine. To Err Is Human: Building a Safer Health System. Washington, DC:National Academy Press; 2000.

Landis NT. Pharmacist on ICU team cuts ordering errors by 66%. Am J Health Syst Pharm 1999;56:1700.

Lapworth R, Teal TK. Laboratory blunders revisited. Ann Clin Biochem 1994;31:78–84.

Leape LL. Striving for perfection. Clin Chem 2002;48:1871–1872.

Leape LL, Brennan TA, Laird N, et al. The nature of adverse events in hospitalized patients. Results of the Harvard medical practice study II. N Engl J Med 1991;324:377–384.

Levey S, Jennings ER. The use of control charts in the clinical laboratory. Am J Clin Pathol 1950;20:1059–1066.

Lord JT. Risk management in pathology and laboratory medicine. Arch Pathol Lab Med 1990; 114:1164–1167.

Lumadue JA, Boyd JS, Ness PM. Adherence to a strict specimen labeling policy decreases the incidence or erroneous blood grouping of blood bank specimens. Transfusion 1997;37:1169.

Mann FA, Danz PL. The night stalker effect: quality improvements with a dedicated night-call rotation. Invest Radiol 1993;28:92–96.

Marks V. False-positive immunoassay results: a multicenter survey of erroneous immunoassay results from assays of 74 analytes in 10 donors from 66 laboratories in seven countries. Clin Chem 2002;48:2008–2016.

McCarthy BD, Beshansky JR, D'Agostino RB, Selker HP. Missed diagnoses of acute myocardial infarction in the emergency department:results from a multicenter study. Ann Emerg Med 1993; 22:579–582.

McIntyre JW, Nelson TM. Application of automated human voice delivery to warning devices in an intensive care unit: a laboratory study. Int J Clin Monit Comput 1989;6:255–262.

McSwiney RR, Woodrow DA. Types of error within a clinical laboratory. J Med Lab Technol 1969;26:340–346.

Morrison AL, Beckmann U, Durie M, Carless R, Gillies DM. The effects of nursing staff inexperience (NSI) on the occurrence of adverse patient experiences in ICUs. Aust Crit Care 2001;14:116–121.

Nevalainen D, Berte L, Kraft C, Leigh E, Picaso L, Morgan T. Evaluating laboratory performance on quality indicators with the six sigma scale. Arch Pathol Lab Med 2000;124:516–519.

Nutting PA, Main DS, Fischer PM, et al. Toward optimal laboratory use. Problems in laboratory testing in primary care. JAMA 1996;275:635–639.

Plebani M, Carraro P. Mistakes in a stat laboratory:types and frequency. Clin Chem 1997; 43:1348–1351.

Pope JH, Aufderheide TP, Ruthazer R, et al. Missed diagnoses of acute cardiac ischemia in the emergency department. N Engl J Med 2000;342:1163–1170.

Pronovost P, Berenholtz S, Dorman T, Lipsett PA, Simmonds T, Haraden C. Improving communication in the ICU using daily goals. J Crit Care 2003;18:71–75.

Pronovost PJ, Dang D, Dorman T, et al. Intensive care unit nurse staffing and the risk for complications after abdominal aortic surgery. Eff Clin Pract. 2001a;4:199–206.

Pronovost PJ, Waters H, Dorman T. Impact of critical care physician workforce for intensive care unit physician staffing. Curr Opin Crit Care 2001b;7:456–459.

Renner SW, Howanitz PJ, Bachner P. Wristband identification error reporting in 712 hospitals. A College of American Pathologists' Q-Probes study of quality issues in transfusion practice. Arch Pathol Lab Med 1993;117:573–577.

Reynolds SL. Missed appendicitis in a pediatric emergency department. Pediatr Emerg Care 1993;9:1–3.

Romanchuk K. The effect of limiting residents' work hours on their surgical training: a Canadian perspective. Acad Med 2004;79:384–385.

Rosen P, Markovchick V, Dracon D. Normative and technical error in the emergency department. J Emerg Med 1983;1:155–160.

Rothrock SG, Green SM, Dobson M, Colucciello SA, Simmons CM. Misdiagnosis of appendicitis in nonpregnant women of childbearing age. J Emerg Med 1995;13:1–8.

Rothrock SG, Skeoch G, Rush JJ, Johnson NE. Clinical features of misdiagnosed appendicitis in children. Ann Emerg Med 1991;20:45–50.

Rusnak RA, Borer JM, Fastow JS. Misdiagnosis of acute appendicitis: common features discovered in cases after litigation. Am J Emerg Med 1994;12:397–402.

Schoenberg R, Sands DZ, Safran C. Making ICU alarms meaningful: a comparison of traditional vs. trend-based algorithms. Proc AMIA Symp 1999:379–383.

Shirm SW, Graham CJ, Seibert JJ, Fiser D, Scholle SH, Dick RM. Clinical effect of a quality assurance system for radiographs in a pediatric emergency department. Pediatr Emerg Care 1995;11:351–354.

Sirota RL. The Institute of Medicine's report on medical error. Implications for pathology. Arch Pathol Lab Med 2000;124:1674–1678.

Stambouly JJ, McLaughlin LL, Mandel FS, Boxer RA. Complications of care in a pediatric intensive care unit: a prospective study. Intensive Care Med 1996;22:1098–1104.

Tarnow-Mordi WO, Hau C, Warden A, Shearer AJ. Hospital mortality in relation to staff workload: a four-year study in an adult intensive-care unit. Lancet 2000;356:185–189.

Thomas EJ, Studdert DM, Burstin HR, et al. Incidence and types of adverse events and negligent care in Utah and Colorado. Med Care 2000;38:261–271.

Tissot E, Cornette C, Demoly P, Jacquet M, Barale F, Capellier G. Medication errors at the administration stage in an intensive care unit. Intensive Care Med 1999;25:353–359.

Urrutia A, Bechini J, Tor J, Olazabal A, Rey-Joly C. Assessment of thoracic X-ray readings by emergency room physicians at a university hospital. Med Clin (Barc) 2001;117:332–333. [in Spanish].

Vincer MJ, Murray JM, Yuill A, Allen AC, Evans JR, Stinson DA. Drug errors and incidents in a neonatal intensive care unit: a quality assurance activity. Am J Dis Child 1989;143:737–740.

Wilson RM, Runciman WB, Gibberd RW, Harrison BT, Newby L, Hamilton JD. The Quality in Australian Health Care Study. Med J Aust 1995;163:458–471.

Wiwanitkit V/. Types and frequency of pre-analytical mistakes in the first Thai ISO 9002: 1994 certified clinical laboratory, a six-month monitoring. BMC Clin Pathol 2001;1:5–9.

Wright D, Mackenzie SJ, Buchan I, Cairns CS, Price LE. Critical incidents in the intensive therapy unit. Lancet 1991;338:676–678.

Zarbo RJ, Jones BA, Friedberg RC, et al. Q-tracks: a College of American Pathologists program of continuous laboratory monitoring and longitudinal tracking. Arch Pathol Lab Med 2002;126:1036–1044.

# 5 Creating a culture for medical error reduction

The issue of medical errors has received substantial attention in recent years. The Institute of Medicine (IOM) report released in 2000 has several implications for health care systems in all disciplines of medicine. Notwithstanding the plethora of available information on the subject, little substantive action has been taken toward reducing medical error. A principal reason for this may be the stigma associated with medical errors. An educational program with a practical, informed, and longitudinal approach offers realistic solutions toward this end. Effective reporting systems need to be developed as a medium of learning from the errors and modifying behaviors appropriately. The presence of a strong leadership supported by organizational commitment is essential in driving these changes. A national, provincial, or territorial quality care council created solely for the purpose of enhancing patient safety and medical error reduction may be formed to oversee these efforts. The bioethical and emotional components associated with medical errors also deserve attention and focus.

## 5.1 Introduction

In recent years, there has been a groundswell of interest from the public, the health care community, and government agencies in addressing the issue of medical errors. It is widely accepted that the present health care system is complex and stressed. This system leads to unsafe conditions for patient care. Reducing the error rates in health care is central to improving overall quality in health care delivery. The health care system's unique needs in quality improvement cannot be met without an effective culture that promotes quality and patient safety. The outcomes in complex work cultures like health care depend mainly on integration of individuals into teams as well as technical and other organizational factors (Bogner, 1994). The reduction of health care errors requires a concerted effort, similar to that undertaken by the commercial aviation and nuclear power sectors. Unless there is an organizational and institutional commitment toward achieving the objective of patient safety, any proposed solutions are futile. To foster a culture of safety, it is necessary to have a culture that supports reporting, justice, flexibility, and learning (Reason, 1997).

The benefits of such a culture are immense. It creates a medical system in which all the stakeholders – patients, medical professionals, and insurers – have increased trust in one another, making confrontations, including costly malpractice claims, less likely to occur. It lessens the stress experienced by physicians, allowing them to take better care of patients. It lowers costs because of the drop in malpractice claims and the reduction in the number of patients who experience extended hospitalization due to adverse events, raising the system's efficiency.

The recent attention and initiatives toward patient safety are encouraging but are not entirely novel. The health care sector has been continuously involved with quality improvement activities, albeit under different names. However, previous efforts have delivered only limited results, and tremendous potential for improvement still exists. A renewed commitment toward quality and safety initiatives is the current need. This poses significant challenges, and not necessarily all the factors are under the control of the profession itself (Blumenthal, 1994). A prime role exists for the accreditation and legislative bodies in promoting patient safety initiatives; patient safety issues should form one of the principles behind the accreditation standards established for all health care sectors. The other contributors to initiatives aimed at promoting patient safety include professional associations, regulatory bodies, health care organizations, and the educational system. In the ensuing sections, we discuss some solutions that form the basic framework for quality improvements in health care sectors that can build on the past efforts of quality improvement and patient safety. We have also reviewed some essential bioethical concerns and the emotional impact of medical errors on health care professionals, as they form an integral part of quality improvement and patient safety initiatives.

## 5.2 Education and professional development

The medical education curriculum is continuously monitored and is being constantly revised in response to ongoing changes in knowledge and practice. There is, however, a lag in developing one critical area of health care, that is, quality care. Given that patient safety is one of the crucial features of quality care, the lackadaisical interest shown by academic authorities on teaching quality and patient safety is quite intriguing. It has persisted even after the IOM proposed the introduction of patient safety issues in the medical education curriculum (Institute of Medicine, 2001). We have previously suggested the need for an educational program at all levels in health care (Kalra and Collard, 2002). We believe that this educational program would serve a definite purpose in enhancing patient safety and reducing medical error. In recent times, some specialties have worked toward developing an educational agenda for teaching about patient safety and error (Croskerry et al., 2000). Researchers have proposed various other curriculum designs for enhancing patient safety (Barach, 2000; Cosby and Croskerry, 2003; Lowery et al., 2001).

A culture in which medical professionals are less tolerant of medical errors must be developed. Discordance between physician perceptions and patient perceptions exists. Hughes (1971) notes that medical professionals are concerned with the actual process of medical care, whereas clients' only concern is the process's end result. This belief stems from the fact that professionals are aware that outcomes are an unreliable criterion to use as the sole basis for assessing performance. As Bosk (2005) writes, this means that "[medical] professionals are much more aware, tolerant, and forgiving of normal error than lay clients." The expectations of medical professionals and their clients ought to be aligned so as to mitigate perceptions of physician inadequacy and complacency on the part of the patient.

The need for an educational program at all levels of health care emerges from one basic necessity, and that is enhanced quality care and patient safety. Qualified physicians and other health care professionals are well prepared with the scientific base of medical knowledge that equips them to provide care for their individual patients, but only a relative few develop the skills that are necessary to continuously strive for quality improvements in everyday practice. Education in this field is essential for health care professionals to further hone their scientific skills and, ultimately, to improve their professional practice. Morbidity and mortality due to health care errors are high, probably higher than they are for some pathological causes of death. These high numbers underline the need for action. Education about these errors and their preventability could provide us with some of the basic interventional approaches to reducing errors and improving quality. This may assume significance when one considers the prevalence of medical mistakes among students and house officers (Estrada et al., 2000; Wu et al., 1993). Wu et al. (1993) reported that 45% of the house officers in internal medicine admitted having made a mistake in practice. The authors, using multivariate analysis of the data, were able to show that house officers who accepted responsibility for their errors were more likely to make constructive changes to their practices than were those who coped by escape or avoidance, who were more likely to report a defensive change in practice. They suggested, among other things, that medical educators provide specific advice about how to prevent reoccurrences of mistakes.

We have suggested that this concept be placed in an organized and systematic format in the learning years through an educational program (Kalra and Collard, 2002). We have previously suggested that quality care grand rounds be instituted as a forum for discussion and evaluation of medical errors in a guilt-free and secure environment aimed at achieving enhanced patient safety (Mulla et al., 2003) (see ▶Fig. 5.1). We have also suggested and implemented a "no-fault" model to encourage voluntary anonymous reporting of all critical incidents by laboratory professionals (Kalra et al., 2003). Our model also includes an educational component that targets all levels of laboratory professionals, including technologists and residents. Briefing and debriefing sessions are held periodically to continuously inculcate the culture of safety and error prevention in laboratory practices. As a part of the process and completion of the loop, the laboratory staff completes a feedback questionnaire at regular intervals that assesses staff attitudes, understanding, and satisfaction with the design of the model and the educational program. The reporting system allows for sufficient scope to review and critically analyze the incident. It accounts for the systemic factors that contribute to errors and deemphasizes individual blame or action. The initiative is supported by the clinical chemistry laboratory professional staff and clearly spells out the management team's attitude toward error reporting and reduction.

To achieve industrial-grade quality in clinical medicine is an overwhelming task. It calls for a total cultural revamp and interdisciplinary cooperation. Satisfactory improvements in achieving a culture of quality and patient safety cannot be achieved without revolutionary changes. There are significant discontinuities in the current health care system, and these need to be minimized through the adoption of an educational approach. The imperative should be to improve and teach these improvements while simultaneously attempting to reform the existing systems. Significant changes can begin by defining a curriculum to inform and educate (Cosby and Croskerry, 2003). An educational program may have various components (see ▶Fig. 5.2); these components need to be emphasized to achieve the maximum benefits from the program.

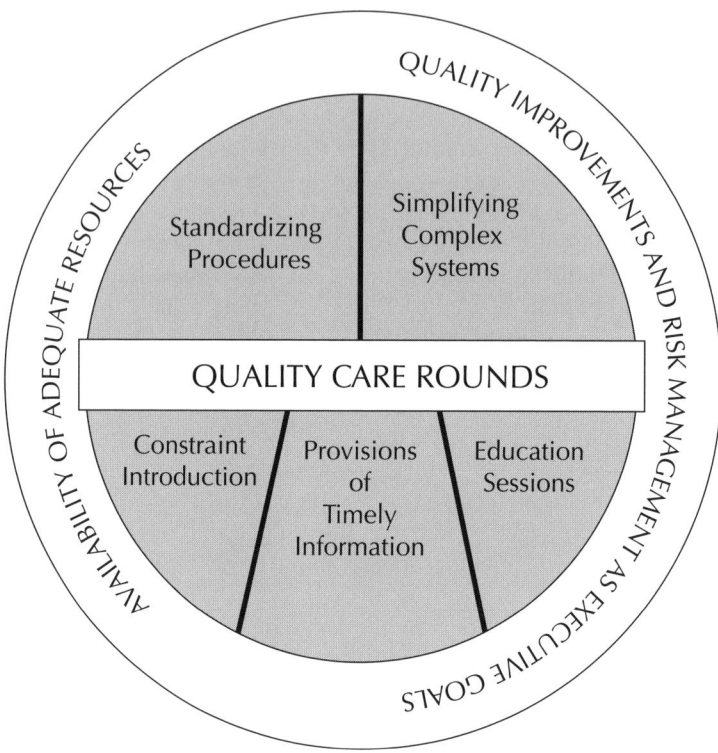

**Fig. 5.1:** Objectives of the quality care rounds.

An educational program should start by creating awareness about the inevitability of error in complex health care systems. Health care professionals need to understand the underlying system and the human issues surrounding an error. Cosby and Croskerry (2003) made this awareness a principal concept in their curriculum for teaching patient safety in emergency medicine. They noted that this awareness and acknowledgment of the reality of error could contribute to a culture that more effectively addresses the inadequacies of the system. The concept of error in every human function is not well accepted or understood by the health care fraternity. Similar views have been advocated earlier. Pilpel et al. (1998) emphasized the need for a teaching program that aims to impart acknowledgment and tolerance of error to undergraduate medical students. The authors suggested that implementation of such a program may defy institutional norms that promote authoritarianism, intolerance of uncertainty, and denial of errors.

The second component of an educational program should focus on achieving a balance between systemic deficiencies and individual responsibility. This attitude helps health care professionals achieve a balance with their working environments and offsets many of the disadvantages offered by the system. Such learning is essential in view of previous research showing that physicians who ascribe errors to systemic factors are less likely to modify their future behaviors and are more likely to repeat these errors again (Wu et al., 1991). The health care professional ought to be taught how to manage errors

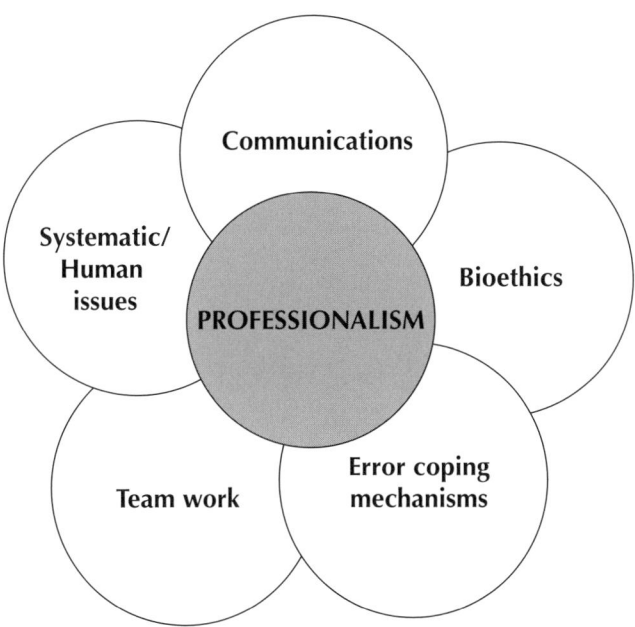

**Fig. 5.2:** Components of an educational program toward medical error reduction.

in a system that is fraught with flaws. This should not underemphasize the systemic influence on error but should emphasize human coping mechanisms that can help minimize the impact of these deficiencies. A balance can be achieved by creating an academic training program that works aggressively to remove certain obstacles that house officers encounter when they look for personal causes of error (Casarett and Helms, 1999).

The third component that needs to be addressed in an educational program relates to teamwork and work ethics. In today's complex medical care, no single individual achieves total patient care. It requires a team effort. The medical care team may consist of one or more physicians, nursing staff, laboratory staff, pharmacists, and medical trainees. Medicine, over the years, has developed a hierarchical structure, with the physician on top, an elite position enjoyed by virtue of his knowledge and academic qualifications (Lester and Tritter, 2001). With no offense intended toward the elite physician, we suggest that this hierarchical structure of the medical care team does not bode well for enhancing patient safety. In such a structure, questioning the professional wisdom of the physician or reporting an obvious incompetence becomes difficult. Health care students have to learn that safe and high-quality patient care is best achieved by small teams, where every member of the team owes the patient the best possible care. A team-oriented approach that allows constructive suggestions from any team member should be encouraged in clinical decision making, and this should be a key point in the educational program. Teamwork in health care systems offers a layered approach through which it becomes difficult for an error to reach the patient. Similar teamwork approaches adopted in the form of crew resource management in aviation have had their desired effect and even convinced skeptics of the value of teamwork and teamwork training (Johnson, 2001). Multidisciplinary teaching that enables medical students to appreciate the roles and skills of other health care professionals has also been emphasized to develop the culture of teamwork (Lester and Tritter, 2001).

Another component of an educational program to reduce health care errors is the development of adequate communication skills. Usherwood (1993) showed the influence of communication skills on clinical consultation behaviors such as active listening and empathy. Good communication skills constitute a significant gain for students and are associated with improvements in patient satisfaction (Novack et al., 1992). Developing these skills is imperative, not only for patient contacts but also for peer interactions and interactions with the family of the patient. The emphasis on teaching communication skills has been clearly underlined in the General Medical Council report *Tomorrow's Doctors* (1993). Other important components that are to be addressed in any educational program include ethical issues and emotional coping after adverse events and errors.

It is essential, too, that development of professionalism be central to the education of medical practitioners. Physicians, for one, have long eschewed the double roles of healer and professional (Cruess and Cruess, 1997). It has been observed that genuine medical professionalism is at peril today (Wynia et al., 1999). The medical curriculum has periodically renewed its focus on developing the role of a health care practitioner as a healer, but it has largely ignored the concept of professionalism. This, however, is changing, and professionalism as a concept is becoming respectable again (Cruess and Cruess, 2000). A renewed commitment toward this end has been seen in the release of a new charter for enhancing professionalism in the new millennium (ABIM Foundation, 2002). It is suggested that the concept of professionalism form an integral component of an educational program to reduce medical errors and quality improvement in health care processes.

There are various options when it comes to teaching about quality improvement and error prevention techniques. It is suggested that, whatever the teaching techniques, education be patient centered and productive. Maximum benefits can be achieved by training health care professionals in life-like simulation exercises, as suggested for anesthesia (Barach, 2000). Gauging the responses of trainees to such situations and training them on the ideal responses should offer a good learning experience for trainees. Other innovative training techniques have also been described. Rall et al. (2002) have reported training techniques that require theoretical medical information as well as practical training sessions and that use realistic patient simulator systems with videotaping and interactive debriefing following the training sessions. Other teaching approaches may include problem-based learning, small-group learning activities, and didactic lectures. Involvement of students in designing the core teaching topics and courses may stimulate motivated and active participation.

This educational program should be implemented at all levels and should not confine itself to undergraduate and postgraduate teaching curriculums. Practitioners also stand to benefit immensely from such a program. The curriculum on quality improvement and patient safety should form a part of continuing medical education. The regulatory and licensing bodies should be actively involved and should mandate adherence. Appropriate scheduling and crediting of sessions should be encouraged. Adequate incentives and resources should be dedicated to promoting periodic education and attendance at the sessions.

## 5.3 Error reporting systems

The genesis of medical errors lies in both system and human violations and failures. In a great proportion of cases, inappropriate reporting markedly inhibits problem solving. These problems with reporting can be mainly attributed to the blame game that commonly follows voluntary reporting. It has been noted time and again that silence, denial, cover-up, or blame, or a combination of these, invariably follows mistakes.

It is imperative for the sake of patient safety and continuous quality improvement in health care that disclosure, reporting, and discussion of errors committed or witnessed be a standard procedure. Only through appropriate disclosure and open attitudes can the cause of mistakes be identified. These mistakes may be in the form of "near misses," actions that almost happened that could have resulted in an adverse event, or "sentinel events," unexpected events resulting in death or serious physical or psychological harm. The study of mistakes is vital to making the changes that can prevent these problems in the future (Casarett and Helms, 1999). If one wishes to treat the errors as learning opportunities, then it is necessary to know about as many errors as possible. This requires a cultural change and sensitive handling of the individual reporting the error (Alberti, 2001). Other well-known methods of analyzing accidents and critical events, such as case reporting, mortality and morbidity conferences, and medical audits, achieve very little in the way of both reporting and substantive follow-up action. The perceived benefits achieved by an effective voluntary or mandatory reporting system may be innumerable.

The term "critical incident reporting" was first used during World War II to describe an objective method of recruiting pilots (Flanagan, 1954). Critical incident reporting is a way of gathering information about critical incidents and events that can be subsequently analyzed to determine the factors that led to the event or events. The technique of critical incident analysis was first introduced into health care by anesthesiologists, who reported significant decreases in mortality rates after the implementation of the technique (Eichhorn, 1989; Pierce Jr., 1996).

The reporting of adverse events and medical errors was one of the principal recommendations of the IOM report *To Err Is Human* (Kohn et al., 2000). The IOM intended that such reporting of critical incidents would increase information sharing and enhance opportunities for learning from these mistakes. This recommendation was controversial and not very well received by the health care fraternity, and understandably so. With no peer review protection or protection from malpractice suits, hospital and health care professionals feared that voluntary reporting systems would lead to personal liability. Apart from these liability issues, there are a number of other factors that lead staff to underreport critical incidents. Cullen et al. (1995) reported that time constraints, fear of punishment, and lack of perceivable benefits were primary reasons for staff's failure to report critical incidents. Vincent et al. (1999) suggested the "blame" culture, high workload and beliefs that the event did not warrant reporting as the reasons for staff's failure to report critical incidents. Another critical component of underreporting remains physicians' attitude toward reporting. Figueiras et al. (1999) suggested that physicians believe that because adverse events associated with certain drugs are already documented by manufacturers, their reporting of such an event would not contribute to medical knowledge. The authors believed that such attitudes among physicians strongly contribute to the failure to report critical incidents. The success of reporting systems depends largely on removal of the disincentives associated with reporting. There is little to be achieved by forcing these reporting systems on health care professionals. Cultural changes that support voluntary reporting must be developed. We are encouraging voluntary reporting in our clinical chemistry laboratory by eliminating some of the known disincentives and offering confidentiality, anonymity, and protection from personal liability.

Many suggestions have been offered to make reporting systems successful (Cohen, 2000; Gaynes et al., 2001; Uribe et al., 2002). There are some vital characteristics that have to be incorporated into the reporting systems if tangible results are to be achieved. Anonymity and confidentiality are the two most important incentives that can be offered to health care professionals to encourage them to report critical incidents.

Nonpunitive reporting is the most important requirement in a successful reporting system. Unless reporters are assured of indemnity from liability resulting from reporting, few will step forward to report critical incidents. This protection is finding increased favor with legislatures and governments (Johnson, 2002).

The second characteristic of the reporting system should be its confidential manner of handling information. Information is provided for the purpose of protecting patients against the possibility that similar harmful events will occur in the future, and it should be used solely for this purpose. Important information is derived from such incident reporting, and revealing the identities of those reporting will only serve to deter future reports. However, ensuring the confidentiality of such information may pose hurdles because of accountability issues. If this is an issue, then sanctions must be limited to serious violations, and public disclosure must withhold details but provide information about the occurrence of the event and the remedial actions to prevent repetition in the future (Leape, 2002). This will provide greater acceptance of voluntary reporting systems among the health care fraternity. Some researchers advocate limited public disclosure of medical errors to health care staff, organizational executives, and patients (Stewart, 2002).

The third characteristic of a successful reporting system has to be its independent nature. This system should not have any role to play in regulatory issues such as licensing. Such a program will reassure those reporting that disclosure will not jeopardize their professional careers and growth. Such systems have been successful in aviation.

The other important principle on which the reporting system should be based on is system-oriented changes, rather than reforms targeted at individuals. The prime objective of reporting should be to gather information to allow the design of more effective preventive strategies, especially against system contributions to such incidents (Bhasale, 1998).

Apart from these four basic principles, an additional requirement for the reporting system is that it provide high-quality information. Quality reporting should augment voluntary incident reports. Liu et al. (2001) reflected on this concern. They reported that a majority of reports of adverse drug reactions did not include adequate information. Other important characteristics of successful reporting systems are the provision of feedback by expert reviewers after they have evaluated the clinical circumstances under which the event took place, timely reporting and dissemination of information, and a responsive program that includes both recommendations from the authority and implementation of recommendations by the participating bodies (Leape, 2002). Reporting without analysis and appropriate follow-up may prove counterproductive and hamper future reporting of critical incidents (Kohn et al., 2000).

The level at which the reporting systems should be implemented and operated is highly debatable. A national reporting system of all adverse events and events averted at the last minute has been advocated for improving patient safety (Kohn et al., 2000). Doubts exist, however, over the feasibility of such a system, in terms of both requirements and economics (Leape, 2002). Local reporting systems, though highly practical and operational, may not serve the purpose. Locally based systems may increase reporting and offer better possibilities for direct feedback, but these may have a smaller database. Another possibility is that local reports may be restricted to only some categories in which errors occur at such low frequencies that generating sufficient information for prophylactic action may be impossible. Under these circumstances, the best compromise may be specialty-restricted reporting systems that allow practitioners to glean the maximum advantage of information from the data generated (Wu et al., 2002). Certain other voluntary systems, such as the Joint Commission on Accreditation of Healthcare Organization (JCAHO) sentinel event reporting program, the MedMARx program of the United States Pharmacopeia, the National Nosocomial Infection Survey of the Centers for Disease Control, and the Institute for Safe Medical Practices' reporting program for the prevention of medication error are already being practiced with success (Leape, 2002).

## 5.4 Leadership and regulatory issues

In order for medical error and patient safety to be addressed in any form, there needs to be appropriate leadership from all levels, including from physicians, organizations, and other related authorities. The IOM report (Kohn et al., 2000) incorporated the concept of leadership into one of its recommendations: "Health care organizations and the professionals affiliated with them should make continually improved patient safety a declared and serious aim by establishing patient safety programs with defined executive responsibility." The IOM report explains that executive responsibility includes providing clear and specific guidelines for others to follow, with incentives to encourage the application of new information.

It is for the leaders to set the standards. O'Leary (2000) argues that leaders in the medical profession and health care organizations should accord high priority to patient safety issues. Acknowledging the limited role of accrediting bodies and regulatory agencies in changing existing attitudes, behaviors, and priorities related to the identification and management of medical errors, the author emphasizes these groups' specific roles in fostering constructive changes in health care organizations. The JCAHO is pioneering efforts in patient safety initiatives, and this tops its list of strategic priorities (JCAHO, 2001a). The JCAHO has recognized the role of leadership in achieving patient safety standards (JCAHO, 2001b). Assuming responsibility is the key for making improvements in the system. Reinertsen (2000) suggested that leaders lead by channeling attention toward our error-prone health care systems.

It is our belief that the leadership role, if taken up by accreditation bodies, could make a significant contribution to a widespread commitment to patient safety. If these organizations mandate patient safety as an accreditation standard, the number of patient safety initiatives will inevitably soar. We have previously suggested that quality and patient safety initiatives should be a part of the accreditation process (Kalra and Collard, 2002).

There are numerous national and local organizations that can assume leadership roles and implement programs; alternatively, a separate commission or institute dedicated solely to the pursuit of quality care and patient safety could be established.

## 5.5 Establishing a quality care council

The case for a quality care council or center has been clearly and appropriately made in the IOM report (Kohn et al., 2000). A key factor in this recommendation by the IOM is the success achieved by similar type of organizations in other sectors, such as aviation.

The quality care council should be set up and dedicated to the sole purpose of quality improvement and patient safety enhancement. It should have a national and regional mandate, with support from both levels. This council should be an independent body free from government or bureaucratic influence, composed of respected individuals well known for their integrity and selected on the basis of their merit, knowledge, expertise, and experience, rather than because they represent any profession, group, or constituency. Furthermore, the quality care council should be independent and should not be linked to any other regulatory, accrediting, or licensing body. Such a governing body would contribute to a public perception of the council as a body established on the basis of genuine principles and safe from unwanted governmental or bureaucratic influence or conflicting interests. Some may argue that if the council is a neutral party and thus not authorized to take corrective action, it may not achieve the desired results in improving quality of care. It is our contention, however, that if the council has authority to punish professionals or organizations at fault, it will only deter voluntary reporting and remedial actions.

This council would set achievable quality care goals, evaluate trends, and implement further actions accordingly. The council could also be responsible for implementing voluntary reporting systems, educational programs, and appropriate research initiatives aimed at enhancing patient safety. It is vital that the council have powers to obtain relevant information from all sources and to suggest recommendatory actions. Such a council would help alleviate the problems created by fragmentation of information by acting as the main body responsible for collection and dissemination of all relevant information. This council could also serve as an interface between health care organizations and government, deriving various concessions from both parties or lobbying for legislative and tort reforms in exchange for the adoption of many patient safety initiatives.

In Canada, there currently exists a Canadian Patient Safety Institute (CPSI), which fills a coordinating and leadership role across the health care sector by promoting patient safety among various stakeholders, such as health care partners, patients and their families, and the general public. CPSI was established in 2003 as an independent, nonprofit body that operates alongside health professionals, regulatory bodies, and governments to advance patient safety within the Canadian health care system. The institute promotes education and professional development to further the knowledge and skills of medical providers, sponsors research to enhance understanding of patient safety, and provides tools and resources to allow health care providers to implement effective governance practices for quality and patient safety (Canadian Patient Safety Institute, 2010).

The CSPI also provides assistance for patients and families, advising patients to ask questions and to talk and listen to their team of healthcare providers to promote communication and health. After all, doctors may fail to ask potentially lifesaving questions or fully clarify medical information to patients. Patients must be aware that many of doctors' personal questions are legitimate, and they too should not feel inhibited about asking questions. The only way to ensure that a doctor or health care provider does not forget to ask important questions is often for patients to ask such questions themselves. CPSI encourages patients, for examples, to ask questions such as "Why are you prescribing this medication, what are the side effects, how will the medication will help me, and when should I take the medication?" (Canadian Patient Safety Institute, 2007). As pharmacist Melanie Rantucci explains, "The patient is important to bringing the community and hospital systems together and in keeping the communication flowing during ambulatory care. The patient is the only common denominator in the process and efforts need to be made to inform and engage patients in medication reconciliation" (Canadian Patient Safety Institute, 2007). This role for the patient must be supported by multifaceted educational and informational promotions by medical professionals themselves and by a public awareness campaign.

## 5.6 Emotional impact of errors on health care professionals

At the other end of the spectrum are health care professionals who have committed medical errors. Mizrahi (1984) classified error coping mechanisms used by medical house staff into three categories: denials, discounting, and distancing. Wu (2000) described health care professionals who have committed an error as the "second victim." In a well-written editorial, Wu noted that, "although patients are the first and obvious victims of medical mistakes, doctors are wounded by the same errors: they are the second victims." Truly, the trauma, dilemma, and emotions faced by health care professionals who are involved with a medical error are severe (Wu et al., 1991). Professionals find solace by disclosing their mistakes or perhaps by confessing their guilt to a trusted colleague who can offer professional reaffirmation or to someone other than their peers (Newman, 1996). It is therefore suggested that adequate support be extended to these other victims of medical error and that standard processes be designed by institutions and professional bodies to deal with the emotions generated by errors committed by health care providers. Apart from physicians, the other members of the health care team, such as nurses, technologists, and pharmacists, who witness an error, are often silent victims. They agonize over their shared loyalties to the patient and to their institution and team (Wu, 2000). However, they have an ethical, if not legal, duty to act (Hebert et al., 2001). Hebert et al. suggested that the witnesses exercise a number of options ranging from encouraging disclosure by the errant professional to discussing the situation with appropriate authorities.

## 5.7 Conclusion

Provision of safe patient care is a major challenge confronting today's health care system. It is imperative that development of a culture of safety be accelerated. Well-designed patient safety initiatives based on systematic interventions may produce the greatest enhancements in the quality of health care processes. These initiatives must be adequately integrated into organizational policies as they are developed. There is also a pressing need for uniform, well-defined policies to address the bioethical component of medical errors, particularly disclosure of errors and the emotional issues attached with them. Providing education to patients to enable them to better take charge of their own health care experience is one such necessary change. Ensuring adequate leadership and developing this leadership are good first steps as we begin to address the gaps in patient safety and change the current cultural climate.

## References

ABIM Foundation. Medical professionalism in the new millennium: a physician charter. Ann Intern Med 2002;136:243–246.

Alberti KG. Medical errors: a common problem. BMJ 2001;322:501–502.

Barach P. Patient safety curriculum. Acad Med 2000;75:551–552.

Bhasale A. The wrong diagnosis: identifying causes of potentially adverse events in general practice using incident monitoring. Fam Pract 1998;15:308–318.

Blumenthal D. Making medical errors into "medical treasures". JAMA 1994;272:1867–1868.

Bogner MS. Human Error in Medicine. Hillsdale: Erlbaum; 1994.

Bosk CL. Continuity and change in the study of medical error – the culture of safety on the shop floor. 2005. http://www.sss.ias.edu/files/papers/paper20.pdf. Accessed July 24, 2010.

Canadian Patient Safety Institute. Annual Review 2010.

Canadian Patient Safety Institute. "Tips for patients and families: ask. Talk. Listen." 2007. http://www.patientsafetyinstitute.ca/English/toolsResources/patientsAndTheirFamilies/Pages/PatientTips.aspx. Accessed July 25, 2010.

Casarett D, Helms C. Systems errors versus physicians' errors: finding the balance in medical education. Acad Med 1999;74:19–22.

Cohen MR. Why error reporting systems should be voluntary. BMJ 2000;320:728–729.

Cosby KS, Croskerry P. Patient safety: a curriculum for teaching patient safety in emergency medicine. Acad Emerg Med 2003;10:69–78.

Croskerry P, Wears RL, Binder LS. Setting the educational agenda and curriculum for error prevention in emergency medicine. Acad Emerg Med 2000;7:1194–1200.

Cruess SR, Cruess RL. Professionalism: a contract between medicine and society. CMAJ 2000;162:668–669.

Cruess SR, Cruess RL. Professionalism must be taught. BMJ 1997; 315:1674–1677.

Cullen DJ, Bates DW, Small SD, Cooper JB, Nemeskal AR, Leape LL. The incident reporting system does not detect adverse drug events: a problem for quality improvement. Jt Comm J Qual Improv 1995;21:541–548.

Eichhorn JH. Prevention of intraoperative anesthesia accidents and related severe injury through safety monitoring. Anesthesiology 1989; 70:572–577.

Estrada CA, Carter J, Brooks C, Jobe AC. Reducing error, improving safety. Medical errors must be discussed during medical education. BMJ 2000;321:507–508.

Figueiras A, Tato F, Fontainas J, Gestal-Otero JJ. Influence of physicians' attitudes on reporting adverse drug events: a case-control study. Med Care 1999;37:809–814.

Flanagan JC. The critical incident technique. Psychol Bull 1954;51:327–358.

Gaynes R, Richards C, Edwards J, et al. Feeding back surveillance data to prevent hospital-acquired infections. Emerg Infect Dis 2001;7:295–298.

General Medical Council. Tomorrow's Doctors: Recommendations on Undergraduate Medical Education. London: GMC; 1993.

Hebert PC, Levin AV, Robertson G. Bioethics for clinicians: 23. Disclosure of medical error. CMAJ 2001;164:509–513.

Hughes EC. "Mistakes at Work" in the Sociological Eye: Selected Papers on Work, Self, and Society. Chicago: Aldine-Atherton; 1971.

Institute of Medicine. Crossing the Quality Chasm: A New Health System for the 21st Century. Washington, DC: National Academy Press; 2001.

Johnson D. How the Atlantic barons learnt teamwork. BMJ 2001;322:563.

Johnson N. Patient Safety Improvement Act of 2002, H.R 4889, 107th Cong.

Joint Commission on Accreditation of Health Care Organizations (JCAHO). Joint Commission president outlines top strategic priorities to aid QI efforts. Hosp Peer Rev 2001a; 26:73–76.

Joint Commission on Accreditation of Health Care Organizations (JCAHO). JCAHO patient safety standards stress leadership. Hosp Case Manag 2001b;9:1–3 [Suppl.].

Kalra J, Collard D. Medical error: a need for an educational program [Abstract]. RCPSC Annals 2002;35:13 [Suppl].

Kalra J, Saxena A, Mulla A, et al. Medical error and patient safety: a model for error reduction in pathology and laboratory medicine [Abstract]. Clin Invest Med 2003;26:15.

Kohn LT, Corrigan JM, Donaldson MS, editors. Committee on Quality of Healthcare in America, Institute of Medicine. To Err Is Human: Building a Safer Health System. Washington, DC: National Academy Press; 2000.

Leape LL. Reporting of adverse events. N Engl J Med 2002;347:1633–1638.

Lester H, Tritter JQ. Medical error: a discussion of the medical construction of error and suggestions for reforms of medical education to decrease error. Med Educ 2001;35:855–861.

Liu BA, Knowles SR, Mittmann N, et al. Reporting of fatal adverse drug reactions. Can J Clin Pharmacol 2001;8:84–88.

Lowery DW, Heilpern KL, Otsuki JA. An interdisciplinary curriculum for reducing medical errors in the academic medical center [Abstract]. Acad Emerg Med 2001;8:586–7.

Mizrahi T. Managing medical mistakes: ideology, insularity and accountability among internists-in-training. Soc Sci Med 1984;19:135–146.

Mulla A, Massey KL, Kalra J. Does every patient benefit from care? Quality care rounds, a patient safety initiative [Abstract]. Clin Invest Med 2003;26:215.

Newman MC. The emotional impact of mistakes on family physicians. Arch Fam Med 1996; 5:71–75.

Novack DH, Dube C, Goldstein MG. Teaching medical interviewing. A basic course on interviewing and the physician-patient relationship. Arch Intern Med 1992; 152:1814–1820.

O'Leary DS. Accreditation's role in reducing medical errors. West J Med 2000;172:357–358.

Pierce Jr. EC. The 34th Rovenstine Lecture. 40 years behind the mask: safety revisited. Anesthesiology 1996;84:965–975.

Pilpel D, Schor R, Benbassat J. Barriers to acceptance of medical error: the case for a teaching program (695). Med Educ 1998;32:3–7.

Rall M, Schaedle B, Zieger J, et al. Innovative training for enhancing patient safety. Safety culture and integrated concepts. Unfallchirurg 2002;105:1033–1042. [in German]

Reason J. Managing the Risks of Organizational Accidents. Aldershot: Ashgate; 1997.

Reinertsen JL. Let's talk about error. BMJ 2000;320:730.

Stewart DW. Conflict in fiduciary duty involving health care error reporting. Medsurg Nurs 2002;11:187–1191.

Uribe CL, Schweikhart SB, Pathak DS, et al. Perceived barriers to medical-error reporting: an exploratory investigation. J Healthc Manag 2002;47:263–279.

Usherwood T. Subjective and behavioural evaluation of the teaching of patient interview skills. Med Educ1993;27:41–47.

Vincent C, Stanhope N, Crowley-Murphy M. Reasons for not reporting adverse incidents: an empirical study. J Eval Clin Pract 1999;5:13–21.

Wu AW. Medical error: the second victim. The doctor who makes the mistake needs help too. BMJ 2000;320:726–727.

Wu AW, Folkman S, McPhee SJ, et al. Do house officers learn from their mistakes? JAMA 1991;265:2089–2094.

Wu AW, Folkman S, McPhee SJ, et al. How house officers cope with their mistakes. West J Med 1993;159:565–569.

Wu AW, Pronovost P, Morlock L. ICU incident reporting systems. J Crit Care 2002;17:86–94.

Wynia MK, Latham SR, Kao AC, et al. Medical professionalism in society. N Engl JMed 1999;341:1612–1616.

# 6 Improving quality in clinical diagnostic laboratories

The majority of errors made in the health care industry are systemic and not individual in nature, but, unless the errors are brought to a supervisor's attention, it will be difficult to correct a flaw in the system. We also recognize that many health care professionals, despite being safe from legal action, will be hesitant to admit errors that could damage their careers and reputations unless the disclosure policy is also to a large extent nonpunitive. This allows physicians to take leadership roles and accept responsibility without fear of undue reprisal. The correction of flawed medical systems and the subsequent protection of patients' health should be the industry's top priority, rather than subjecting physicians who make inevitable errors to punitive measures. Inappropriate blame attribution to serve regulatory needs will merely alienate professionals and discourage them from participating in system improvement.

## 6.1 Introduction

Clinical diagnostic laboratories play a critical role in the diagnosis of many human diseases; they have been doing so for decades now (Kotlartz, 1998). Laboratory testing influences a majority of clinical decision making. With laboratories having such a high degree of influence, it is easy to see the importance of quality in laboratory testing. In today's health care environment of managed care and cost-containment processes, laboratorians have to work collaboratively with other health care professionals, with the sole focus being improvement in medical outcomes. It has been suggested that the importance of laboratory scientists must be proven in order to guarantee the quality of tests and thereby improve the quality of laboratory services (Plebani, 1999). In a health care delivery system that is interdependent with other departments, where the quality of one department has an effect on another, it becomes essential for clinical diagnostic laboratories to set high standards for other departments to follow.

The solution to medical error requires a systemwide overhaul, which should also address the tendency to blame individuals' actions. Yet, when errors do occur, procedures must be in place to ensure that valuable information about what went wrong can be used to make processes safer, instead of being concealed for fear of liability. Doing so would require a system of assurance for health professionals, covering both educational and legal aspects. Errors continue to happen in all three stages of analysis and include human and technical factors. Already, proficiency testing agencies, one of the best external quality assessment methods, are delivering standards for accuracy, though all assessment methods are not of equal value and all have limitations. We will first address the development of quality management programs and the extent of error in clinical diagnostic laboratories, followed by a discussion of the other quality assurance systems; finally, we will propose a no-fault model for clinical laboratories.

## 6.2 Efforts and programs to ensure quality in clinical diagnostic laboratories

Statistical quality control (QC) was first introduced in clinical laboratories by Levey and Jennings in 1950 (Levey and Jennings, 1950). QC gained wide acceptance in later years, and most laboratories adopted it as a standard of practice by the 1960s. Laboratory medicine has been at the forefront of many quality improvement initiatives since then. It has been demonstrated previously that modern quality tools and techniques have been applied to improve medical processes by finding the causes as well as solutions to the defects plaguing the system (Plebani and Carraro, 1997; Witte, 1997).

In clinical diagnostic laboratories, mistakes and blunders contribute primarily to erroneous laboratory results. The precise magnitude of the error rate is difficult to determine for two important reasons: underreporting or a complete lack of feedback and the difficulty of detecting errors. The standard practice followed by laboratories is to report quality indicator data as a percent variance. This most often yields very low values and leaves the laboratories with an exaggeratedly favorable idea of the quality of their performance. Despite the low error rates, the magnitude of usage of clinical laboratories in health care is so high that even the low variances translate into a very high number of defects, as a small percentage of a big number can itself be a big number (Nevalainen et al., 2000). It is in such instances that the true value of adopting six-sigma quality initiatives can be appreciated. It should also be noted that in the six-sigma methodology, errors are expressed as rates and not as absolute numbers.

The research on error and blunder rates in clinical diagnostic laboratories is scarce. However, the few studies that have been reported give varying results. McSwiney and Woodrow (1969) reported a blunder rate of 2% to 3%. This was followed by another study that detected a blunder rate of 0.3% in a large clinical biochemistry laboratory (Chambers et al., 1986). Kazmierczak and Catrou (1993) reported a total error rate of 9.36% in their study of 438 results of replicate creatinine analysis. Lapworth and Teal reported a blunder rate of less than 0.1% of requests in their study of two district laboratories in the United Kingdom, which lasted more than a year (1994). The study reported that nearly 120 blunders were committed in a total of approximately 1 million test results, which translated into approximately 120 defects per million (DPM) opportunities (see ▶Tab. 6.1). In Australian chemical pathology laboratories, the reported error rates have been as high as 39% for transcription and 26% for analytical results, with the best laboratory performing error-free business only 95% of the time (Khoury et al., 1996). This high rate of laboratory errors is much worse than the predicted 16.6% rate of adverse events that has been reported in hospital admissions in Australia (Wilson et al., 1995). Witte et al. (1997) reported 447 DPM unacceptable results and suggested that these were a result of special-cause variation. The authors suggested that the results, which were likely to alter patient care, occurred at a rate of 41 DPM opportunities. Another study on errors in a stat laboratory revealed a relative frequency of 0.47% over three months at various departments of a university hospital (Plebani et al., 1997). An error rate of 0.38% was reported in clinical genetic testing laboratories during a 10-year period (Hofgartner et al., 1999). In a Thai clinical laboratory with ISO 9002:1994 certification at a large hospital, a total error rate of 0.13% was detected (Wiwanitkit, 2001). Bonini et al. (2002) studied the laboratory testing error rates in inpatients and outpatients and reported an error rate of 0.60% and 0.039%, respectively. The authors attributed the large difference in the two settings primarily to the lower skill level of ward staff in blood drawing, the greater complexity of tests performed, and the higher frequency of blood drawings for inpatients.

More recently, Ismail et al. (2002) reported a total error rate of 0.53% in the analytical phase of common immunoassay tests for thyroid stimulating hormones and gonadotropins. The authors concluded that these errors were a result of analytical interference and stressed the importance of early interference in cases where results were not compatible with the clinical scenario. Marks (2002) studied the influence of analytical interference on assays of 74 analytes. A total of 66 laboratories across seven countries participated in the study. Marks found that 8.7% of the results were erroneous, and 49% of these erroneous results were not corrected even by the addition of a blocking reagent.

**Tab. 6.1:** Review of error rates in clinical diagnostic laboratories represented in DPM opportunities.

| Authors | Study Design | Total Analyte/ Results | Error Rates[1] | DPM Opportunities | Data Collection Period |
|---|---|---|---|---|---|
| Chambers et al., 1986 | Prospective | Data not available | 0.3% | NA | 9 weeks |
| Kazmierczak and Catrou, 1993 | Prospective | 438 | 9.36% | 93,607 | 8 days |
| Lapworth and Teal, 1994 | Prospective | 998,018 | < 0.1% | 120 | 1 year |
| Plebani and Carraro, 1997 | Prospective | 40,490 | 0.47% | 4,668 | 3 months |
| Witte et al., 1997 | Retrospective | 219,353 | 0.08% | 807 | 18 years 6 months |
| Hofgartner and Tait, 1999 | Retrospective | 4,234[2] 88,394[3] | 0.38%[2] 0.33%[3] | 3,779[2] 3,337[3] | 10 years[2] 1 year[3] |
| Wiwanitkit, 2001 | Prospective | 941,902 | 0.13% | 1,316 | 6 months |
| Marks, 2002 | Prospective | 3445 | 8.7% | 87,083 | NA |
| Ismail et al., 2002 | Prospective | 5310 | 0.53% | 5,273 | NA |

Notes:

NA = Not Applicable.

[1] Different authors have used different terminologies (i.e., errors, mistakes, blunders, outliers, problems, unacceptable results).

[2] Inspected laboratories.

[3] Data collected through a survey questionnaire.

Clinicians overwhelmingly rely on data generated by laboratory to aid in their clinical decisions. They assume that laboratorians, being trained in QC and quality assurance, detect errors before the reports leave the laboratory. Identification of random errors such as interferences occurring in a laboratory is difficult, but failure to find such errors may affect patient care; Marks (2002) advised clinicians to be aware of these limitations of laboratories. Clear and direct communication between laboratorians and clinicians regarding doubtful or clinically suspect results may be a simple and safe way of providing quality care to the patients.

As seen in Table 6.1, studies of the errors in clinical diagnostic laboratories give a wide range of error rates, varying from 0.1% to 9.36% (Kazmierczak et al., 1993), with an Australian study reporting an error rate as high as 39% for transcription errors alone (Khoury et al., 1996). Various factors may account for this difference in the rates of error cited by the different studies. It may be a result of the different study designs – prospective and retrospective – adopted. A second reason may be the variability of the process itself at the specific time of the study, with some laboratories performing at their peak quality level and others performing at their worst levels. A third reason may be the different criteria adopted by authors to define error, with some authors being very strict in their definition and others being relatively lenient. A fourth factor may be the differences in the methods for choosing a laboratory to be studied, with some laboratories having higher quality standards than others, as a result of which some chosen laboratories may be performing exceptionally with regard to quality, while others may be underperforming. A fifth explanation may be laboratories' use of imperfect error detection methods that allow some errors to go unreported. Whatever the limitations of the reported data, the fact remains that clinical diagnostic laboratories are error prone, and abundant opportunities for improvements in the process exist. These improvements may translate into beneficial outcomes to the patients.

A standard laboratory process is usually divided into three stages: preanalytical, analytical, and postanalytical. An error at any step during the acquisition, processing, and analysis of a specimen and the reporting of a laboratory result can invalidate the quality of analysis and cause the laboratory to fall short of its quality goals (Westgard et al., 1987). The types of errors detected in clinical laboratory services are similar in the United Kingdom and the United States (Grannis et al., 1972; McSwiney et al., 1969). The geographical similarities apart, a majority of the literature on error rates in clinical laboratories has agreed on one other point: the analytical stage of clinical laboratories is more efficient than the other two stages and leagues ahead in quality performance. Lapworth and Teal (1994) state that approximately 32% of total errors occur in the analytical stage, mainly because of wrong patient sample analysis.

Khoury et al. (1996) studied the rate of transcription and analytical errors in Australian chemical pathology laboratories and reported error rates as high as 39% in the transcription of reports and as high as 26% for analytical results in the worst-performing laboratory. Plebani and Carraro (1997) estimated that a huge 68.2% of errors occurred in the preanalytical stage of testing, 18.5% in the postanalytical stage, and 13.3% in the analytical stage of a laboratory process. In the clinical genetic testing laboratories, 60% of errors occurred in the preanalytical phase, 32% in the analytical phase, and a mere 8% in the postanalytical phase (Wiwanitkit, 2001). However, a study by Kazmierczak and Catrou (1993) presents a different view. The authors suggested that a phenomenal 95% of the errors in the study could be attributed to the analytical stage of testing, though the criteria they used to classify errors could be termed rather strict. Wiwanitkit studied the types and frequency of preanalytical errors in a large hospital laboratory and found that approximately 85% of errors occurred in the preanalytical stage, whereas a mere 4.35% of errors occurred in the analytical stage.

The quality cycle in a laboratory is not dependent on the control of analytical processes alone. The precision and accuracy of reported laboratory results are also dependent on the accuracy of the preanalytical and postanalytical stages of the testing process. The high rates of pre- and postanalytical errors necessitate the involvement of nonlaboratory personnel, including clinicians, in order to improve the quality of laboratory results. The heavy balance of errors occurring in the pre- and postanalytical stages of a laboratory testing process reconfirms the susceptibility of the analytical process to human error. It has been estimated that up to 97% of mistakes occurring in laboratory processes result from human error (Goldschmidt et al., 1995). It is therefore suggested that clinical laboratories employ automation and robotics and minimize human involvement in the process wherever possible. Laboratory automation provides for standardized workflow and helps eliminate many error-prone steps undertaken by humans. In doing so, it provides an opportunity for processes to reduce the influence of human factors such as stress, fatigue, negligence, and cognitive impairment. In addition, it enhances the quality of laboratory results and reduces the turnaround times for results.

## 6.3 Proficiency testing in clinical laboratories

Laboratory results play a major role in guiding decisions concerning patient management. As such, it is imperative that lab results be accurate so that clinicians can determine the proper course of medical action. After all, clinical specimens represent valuable medical information about someone's parent, child, sibling, or spouse. Despite the level of automation found in today's clinical diagnostic laboratories, specimens are still mislabeled, misplaced, and misinterpreted. Most serious of all, perhaps, it also happens that the wrong patient's results are used to determine another's medical treatment. Inaccurate results can cause over- or undertreatment or even no treatment at all. Many strategies have been adopted to reduce lab errors, including internal quality control (QC) procedures, external quality assessment (EQA ) programs, certification of education programs, licensing of laboratory professionals, accreditation of clinical laboratories, and the regulation of laboratory services.

Despite efforts to reduce errors in clinical laboratories via the establishment of international standards that harmonize laboratory practices, errors continue to occur. EQA programs are used to address this issue of quality in the laboratory. The principal goal of EQA is to determine the acceptability of laboratory results by evaluating their harmonization with reference data. Proficiency testing (PT) is an external assessment of a laboratory's analytical performance in comparison to that of its peers or to an accuracy-based reference system. PT serves as a regulatory process, whereas EQA typically addresses self-assessment and improvement (Miller, 2003). The majority of EQA programs, including PT, use conventional processed materials to evaluate participants by comparing laboratory results to those of peer laboratories. This allows individual laboratories to determine if they are applying a measurement technology correctly and achieve results in agreement with those of other laboratories.

In order to properly screen and diagnose patients, individual results must be referenced against values or cut-off values. Any purely analytical results do not provide all the required information. To monitor patient situations, however, the follow-up of results can be accomplished only via analytic and intra-individual information. This turnaround time is an important factor in patient care.

## 6.4 External quality assessment and proficiency testing programs

PT is one of the most efficient EQA approaches. In EQA , numerous laboratories report their result to a centralized agency for evaluation. Each participating laboratory then receives a report of its performance alongside a summary of data for other participating laboratories that achieved satisfactory results in a peer group format. Satisfactory performance, according to the Clinical Laboratory Improvement Act (CLIA), for example, requires a score of 80% or better on each challenge in either two consecutive tests or two of three tests within the span of one year. The results must be reported within the time frame specified by the instructions that come with the samples. Failure to submit results within the required time frame results in a score of zero. According to US regulations, it is stipulated that "the samples must be examined or tested with the laboratory's regular patient workload  by personnel who routinely perform the testing in the laboratory, using the laboratory's routine methods" (US Code of Federal Regulations 493.801, 2010).

PT agencies believe that comparison to the most relevant instrument/reagent combinations offers the most accurate assessment of a laboratory's performance. Passing the proficiency testing is a condition for accreditation and licensing  of a clinical laboratory under various regulations (Clinical Laboratory Improvement Act, 1988). In Canada, it is required by provincial regulatory bodies – in the case of Saskatchewan, the College of Physicians and Surgeons of Saskatchewan, or CPSS – that each regional or hospital laboratory enroll in an approved PT program. A central regional laboratory operates this process.

EQA programs should have four elements (Libeer, 2001). First, they should have a participation-based performance method of evaluation. In an ideal scenario, the identity of the control samples should remain unknown to the laboratory team, allowing them to be evaluated as part of daily routine practice. It is suggested that proper performance evaluation be subjected to professional judgment (Libeer, 2001). Moreover, in addition to analytical results, immunization status, clinical interpretation, and suggestions for diagnosis, as well as additional laboratory investigations, should also be reviewed.

Second, EQA  programs should have a method-based performance evaluation. Comparing results obtained by users of the same in-vitro diagnostics can reveal differences in how labs go about their analyses (Libeer, 2001). Third, post-marker vigilance should exist in EQA programs to ensure that accredited laboratories continue to achieve desired results. Finally, EQA programs should provide training and offer help to prospective laboratories who are applying for accreditation (Libeer, 2001). Fahey et al. (2000) maintain that many of the advantages that new, inexperienced, or otherwise disadvantaged laboratories derive from participating in EQA programs result from the fact that they are able to consult with more experienced personnel. PT programs typically provide participants, manufacturers, and standardization organizations with informa-tion about such things as the traceability of methods to reference systems, the effec-tiveness of a manufacturer's transfer of calibration traceability to routine measurement procedures in the field, and the degree of harmonization among routine measurement procedures from different manufacturers.

Peer grouping and analysis of PT samples using a reference measurement are the two most common procedures for assigning target values. Another approach is to base EQA target values upon "true values," which are determined with definitive or reference methods (Stöckl et al., 1993). Such values are designed to reflect an accuracy base in clinical chemistry measurements, transferring the accuracy base to the routine laborato-ries by the direct validation of routine methods against reference methods – preferably in the state of method development – and by the assessment of routine methods in EQA schemes operating with accuracy-based target values (Stöckl et al., 1993).

It is important to note that limitations exist in the analysis of PT results. The majority of sample materials used in PT assessments are modified during preparation in such a manner that they are no longer commutable with the native clinical samples. In these cases, peer grouping is important to ensure that different laboratories are employing similar measurement conditions so that the results are comparable. In special cases where the PT samples are commutable with the native clinical samples, use of a refer-ence measurement procedure is applied, instead. Unacceptable or incongruent results require an investigation to identify the root cause of the error so that corrective action to rectify the situation may be taken. Such an investigation is documented, with a given cause.

Spath identifies the seven steps involved in performing a root cause analysis of medical error (1998). First, people are selected for the investigation team. Second, these individuals determine the sequence of events that led up to the medical error. Third, the event's causal factors are identified. Fourth, from these factors, the root, or most important, causes must be selected. Fifth, the investigation team or another appropriate committee develops and implements corrective actions. Sixth, the com-mittee reports its suggestions to hospital authorities. The seventh and final step is an evaluation of the effectiveness of the actions undertaken. Root cause analyses have been performed for various types of events, including "near misses" (Berry and Krizek, 2000), adverse drug events (Rex et al., 2000), and unsuccessful blood transfusions (Spath, 1999).

If nothing can be indentified to explain an error, it is considered to be a random event. In such a case, the result may be due to a nonpersistent random event. Cor-rective action is not appropriate in such an instance, because the corrective action, since it is not based on clear evidence, might result in additional errors. As a result, it is common practice to perform repeated measurements of PT samples that yield unac-ceptable results or on other samples from the same set to confirm that the original result was indeed a random event. When multiple test results yield relatively large differences that are scattered on either side of the target value, it suggests that the method used lacks adequate precision. Multiple results with relatively large differences in the same direction, however, suggest a bias problem.

The overall participants' mean is the traditional method for setting target values in EQA. These peer-group measurements are obtained by adding a specific amount of analyte into a matrix that is free from the specific analyte. As a result, the quality of the control materials used becomes extremely important. Various PT programs use commutable samples, which are freshly collected and minimally processed human samples.

## 6.5 "No-fault" model

Our no-fault model can address that which EQA alone cannot, as it has an ability to confront error (see ▶Fig. 6.1). While designed primarily for clinical laboratories, it can be adjusted for other spheres of the health care industry. Whenever an error occurs, an error incident report is made voluntarily by anyone who witnessed the error. Those

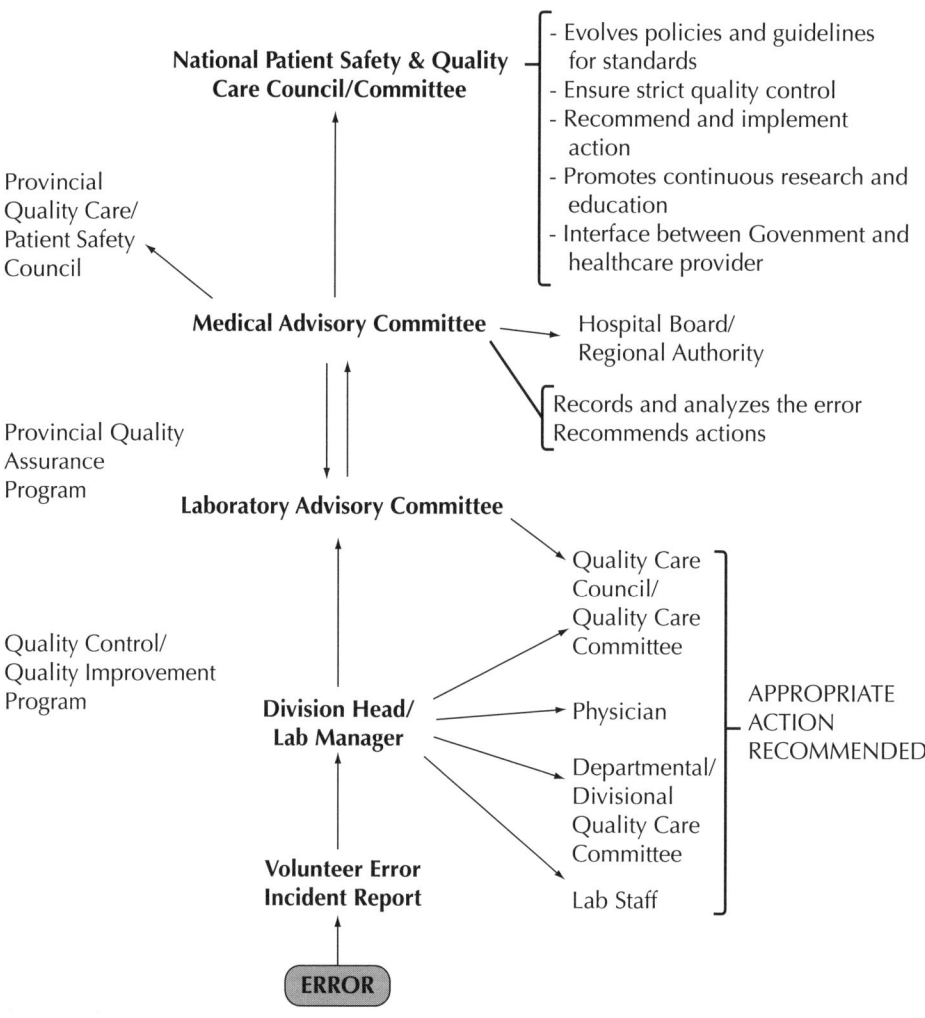

**Fig. 6.1:** No-fault model for error reduction and patient safety in laboratories.

involved in the error are not subjected to punitive measures (expectations will be made in special cases of gross negligence, as when a health care professional's exceptional failure to use reasonable care results in damage or injury to another, and in these cases the error report will be kept confidential).

The report is disclosed to the division head, who takes the appropriate action with laboratory staff, physicians, and the quality care committees that were previously discussed in this chapter. The division head then contacts the laboratory advisory committee, which remains in communication with the quality care committee, as well as the medical advisory committee. The medical advisory committee records and analyses the error and then recommends actions both downwards, to the laboratory advisory committee and the national patient safety and quality care council, and upwards, to the regional quality care council, the hospital board or appropriate regional authority, and the national patient safety and quality care council committee. This final committee is responsible for acting as an interface between the government and health care providers. It promotes research and education, offers evolving policies and guidelines for quality control standards, and recommends and implements action. The Canadian Patient Safety Institute, discussed in chapter 5, performs a job similar to the proposed national council.

Such a model necessitates the integration of certain laboratory activities. First, it requires collaboration between laboratories and industry, professional, accreditation, and regulatory bodies. Second, it calls for clinical laboratory science professional education in error prevention and reporting. Third, it depends on clear and effective communication practices among the all members of the laboratory. Moreover, it requires an emphasis on increased automation and robotics in order to reduce human error. Ultimately, in addition to a model that can address fault and blame, proficiency testing should take a central role in laboratories.

## 6.6 Conclusion

Clinical laboratories should be able to attain peak performance levels in terms of both medical error reduction and cost effectiveness. This requires modern quality management techniques focused on enhancing patient safety. Management must adopt an approach of designing safer systems, evaluating the success and benefits of the systems via proficiency testing programs, and practicing these systems efficiently. Quality assurance management programs, which found their footing in clinical diagnostic laboratories, have developed and improved since the 1950s. Yet, studies continue to show that great numbers of errors are occurring, for a variety of reasons, in all stages of the analysis process. Programs such as EQA and PT are promising starts to achieve greater safety, but they can be much improved with the addition of no-fault programs that better deal with errors if they do occur.

## References

Berry K, Krizek B. Root cause analysis in response to a "near miss." J Healthc Qual 2000;22:2–12.

Bonini P, Plebani M, Ceriotti F, Rubboli F. Errors in laboratory medicine. Clin Chem 2002;48:691–698.

Chambers AM, Elder J, O'Reilly DS. The blunder rate in a clinical biochemistry service. Ann Clin Biochem 1986;23:470–473.

Clinical Laboratory Improvement Act, 1988. Association of American Physicians and Surgeons. http://www.aapsonline.org/msas/clia.php. Accessed July 26, 2010.

Fahey JL, Aziz N, Spritzler J, et al. Need for external proficiency testing program for cytokines, chemokines, and plasma markers of immune activation. Clin Diag Lab Immun 2000;7:540–548.

Goldschmidt HMJ, Lent RW. Gross errors and workflow analysis in the clinical laboratory. Klin Biochem Metab 1995;3:131–140.

Grannis GF, Grumer HD, Lott JA, Edison JA, McCabe WC. Proficiency evaluation of clinical chemistry laboratories. Clin Chem 1972;18:222–236.

Hofgartner WT, Tait JF. Frequency of problems during clinical molecular-genetic testing. Am J Clin Pathol 1999;112:14–21.

Ismail AA, Walker PL, Barth JH, Lewandowski KC, Jones R, Burr WA. Wrong biochemistry results: two case reports and observational study in 5,310 patients on potentially misleading thyroid-stimulating hormone and gonadotropin immunoassay results. Clin Chem 2002;48:2023–2029.

Kazmierczak SC, Catrou PG. Laboratory error undetectable by customary quality control/quality assurance monitors. Arch Pathol Lab Med 1993;117:714–718.

Khoury M, Burnett L, Mackay MA. Error rates in Australian chemical pathology laboratories. Med J Aust 1996;165:128–130.

Kotlarz VR. Tracing our roots: the broadening horizons of clinical laboratory practice (1945–62). Clin Lab Sci 1998;11:339–345.

Lapworth R, Teal TK. Laboratory blunders revisited. Ann Clin Biochem 1994;31:78–84.

Levey S, Jennings ER. The use of control charts in the clinical laboratory. Am J Clin Pathol 1950; 20:1059–1066.

Libeer JC. Role of external quality assurance schemes in assessing and improving quality in medical laboratories. Clin Chim Acta 2001;309:173–177.

Marks V. False-positive immunoassay results: a multicenter survey of erroneous immunoassay results from assays of 74 analytes in 10 donors from 66 laboratories in seven countries. Clin Chem 2002;48:2008–2016.

McSwiney RR, Woodrow DA. Types of error within a clinical laboratory. J Med Lab Technol 1969;26:340–346.

Miller WG. Specimen materials, target values and commutability for external quality assessment (proficiency testing) schemes. Clinica Chimica Acta 2003;327:25–37.

Nevalainen D, Berte L, Kraft C, Leigh E, Picaso L, Morgan T. Evaluating laboratory performance on quality indicators with the six sigma scale. Arch Pathol Lab Med 2000;124:516–519.

Plebani M. The changing face of clinical laboratories. Clin Chem Lab Med 1999;37:711–717.

Plebani M, Carraro P. Mistakes in a stat laboratory: types and frequency. Clin Chem 1997;43:1348–1351.

Rex JH, Turnbull JE, Allen SJ, vande Voorder K, Luther K. Systemic root cause analysis of adverse drug events in an tertiary referral hospital. J Qual Improv 2000;26:563–575.

Spath P. Medical errors: root cause analysis. Or Manager 1998; 14(9):38–41.

Spath P. Your RCA: how do you select events for analysis? Hosp Peer Rev 1999:97–99.

Stöckl D, Reinauer H. Candidate reference methods for determining target values for cholesterol, creatinine, uric acid, and glucose in external quality assessment and internal accuracy control. Clin Chem 1993;39:993–1000.

US Code of Federal Regulations. Public Health-Laboratory Requirements, Title 42, Part 493.801. Condition: enrollment and testing of samples, laboratory requirements (Subpart H), 2010. http://ecfr.gpoaccess.gov/cgi/t/text. Accessed July 26, 2010.

Westgard JO, Klee GG. Quality Assurance. In: Tietz NW, ed. Fundamentals of Clinical Chemistry. 3rd ed. Philadelphia: W. B. Saunders; 1987:238–253.

Wilson RM, Runciman WB, Gibberd RW, Harrison BT, Newby L, Hamilton JD. The Quality in Australian Health Care Study. Med J Aust 1995;163:458–471.

Witte DL, VanNess SA, Angstadt DS, Pennell BJ. Errors, mistakes, blunders, outliers, or unacceptable results: how many? Clin Chem 1997;43:1352–1356.

Wiwanitkit V. Types and frequency of pre-analytical mistakes in the first Thai ISO 9002: 1994 certified clinical laboratory, a 6-month monitoring. BMC Clin Pathol 2001;1:5–9.

# 7 Barriers to open disclosure

Disclosure of adverse events is critical in health care. In the case of a bedridden patient whose body is undergoing heavy drug treatment to resist an aggressive disease or that of a patient under anesthesia, it is extremely difficult for staff to perceive whether or not some particular medical intervention has resulted in personal injury. This is a distinct difference between health care situations and those in other high-risk sectors, such as aviation or other transportation modalities. After all, it is quite easy to detect when an error on a cab driver's part has led to a personal injury. Because of the complexity of the health care environment, addressing potential error requires that we overcome restrictive barriers to disclosure that can and do undermine individual feelings or efforts to practice in an ethically sound manner.

## 7.1 Introduction

In this chapter we offer recommendations for how to initiate and successfully utilize disclosure to improve communication while minimizing negative legal or liability repercussions. Failure to disclose adverse events to a patient harms patients in two distinct ways. First, nondisclosure compromises the patient's ability to undergo corrective medical treatment to mitigate the error. If error is not disclosed to patients in a timely fashion, certain injuries sustained due to medical error may no longer be reversible, resulting in permanent disability or even death. Second, it means that patients are unaware of compensation that they may be entitled to, whether through civil action or a compensation package designed by health services for such an event. Patients should not have to pay full fees for medical treatment gone awry, nor should they have to pay for corrective action necessitated by medical error. Nondisclosure may force patients to do both.

## 7.2 How to disclose

The individual who should disclose the information to the patient is the one most appropriate to handle the discussion, as identified by the concurrence of the medical team that has worked with the patient. This might be a doctor, nurse, hospital manager, or social worker. The appropriate time to disclose largely depends upon the discretion of the discloser. As a general rule, however, disclosure should be made as promptly as possible, while taking into account the patient's medical and emotional condition.

A hospital in Winnipeg, Canada, for example, disclosed errors committed by a senior pathologist who misinterpreted tissue samples, causing two patients with cancerous tumors to be told that they did not have cancer. A review of the pathologist's work detected the errors and concluded that those two patients, one with colon cancer and one with thyroid cancer, received the wrong diagnosis because of the pathologist's errors. The review also found that the pathologist made errors in 40 other pathology reports, or around 5% of the examined cases, although the other mistakes did not impact patient treatment. Both affected patients were immediately notified, and parts of the completed report were later made public (Canadian Broadcasting Corporation, 2008).

Disclosure is a process; the discloser must avoid speculating as to how the error occurred and simply state what is known at the time. The nature, severity, and cause of the unanticipated outcome should be presented in a straightforward and nonjudgmental manner: individual blame should not be meted out. The discloser should express regret about the unanticipated result on behalf of the entire medical team and should, moreover, educate the patient, as well as the patient's family, about the clinical implications of the unanticipated outcome and the medical team's plan for the patient's future medical care. Any questions or concerns put forth by the patient should be answered to the best of the discloser's ability.

It is also important that the disclosure be followed up with a second conversation with the patient, preferably conducted by the same individual who performed the original disclosure. The patient should be given all available information on the adverse event investigation and told of the progress of any new medical treatment. The patient and family should be offered an opportunity to discuss the issue with other relevant health care professionals, who can offer additional information as well as a second opinion on the matter. Moreover, all previously queries left unanswered because of a lack of contextual knowledge on the discloser's part should be responded to.

## 7.3 Disclosing errors to multiple patients

Many adverse events affect multiple patients. Faulty medical equipment in a laboratory continuously rendered inaccurate results for hormone receptor over more than 10 years, resulting in a lack of treatment for many patients who should have tested positively. Beverley Green was one such patient. She found a lump in her breast but tested negatively for hormone receptors, meaning that she wasn't eligible for the hormone therapy Tamoxifen, which can drastically reduce rates of cancer recurrence. Since then, cancer has spread to her liver. Although the province's Eastern Health authority detected the errors in 2005, many patients weren't informed until months or years later. More than 100 patients who were incorrectly diagnosed have since died (Breen, 2009). Disclosing errors to multiple patients is extremely challenging for health care authorities, and few guidelines for disclosure exist that go beyond the individual to the systemwide level.

The nature of a large-scale medical error presents medical administrators with a unique barrier to disclosure. Many large-scale errors involve multiple medical departments within a hospital, as well as multiple organizations. Under current tort culture, where class-action lawsuits are the norm whenever large-scale errors occur, the concerned departments and organizations may be unwilling to cooperate and share information out of fear that such actions will be used to prove their department's liability in court. This, in turn, lessens the chance that timely disclosure and preventative action for affected patients will occur.

In order to minimize the potential harm done by medical errors on a large scale, as soon as an error is identified (which can prove challenging, as many errors as repeated precisely because they are difficult to identify), a review must be conducted that identifies all patients whose care has potentially been jeopardized. This review must determine whether inaccurate diagnoses or subsequent erroneous treatments occurred because of the error; if so, then physicians must take appropriate measures to rectify the patients' care as soon as possible. Patients and other stakeholders (e.g., affected hospital personnel, external authorities, insurers, external health care institutions, and the public) must be informed as soon as possible. This must be done even before a study of the extent and full impact of the error is completed.

Since the error must be disclosed before it is fully resolved, disclosure of large-scale medical errors must be an ongoing process. Patients, stakeholders, and the public must be updated as investigation of the error continues, and the investigative process must be as transparent as possible. The focus of such an investigation must be to ensure quality of care for affected patients, rather than to punitively identify involved physicians. Such punitive behavior will merely serve to alienate professionals and concerned medical institutions, as large-scale errors frequently involve multiple hospitals and clinics. Instead, institutions should avoid playing the "blame game" and cooperate to ensure quality of care. Doling out blame will only prove to be an obstacle for delivering the required health care to the affected patients.

Once the large-scale error has been identified, the project team is created. This is an independent team that, through a transparent process, will determine which patients' care was jeopardized by the error. The project team must inform the medical staff, affected patients, and the public as soon as possible, even before the extent of the error is fully known. The medical staff must provide treatment to those patients adversely afflicted in order to rectify the situation, possibly giving them priority over other patients. This type of practice is only starting to occur.

## 7.4 Bioethical viewpoints

Health care errors continue to be abundant; their seriousness varies from one error to another. Sometimes the cause may be faulty systems, and at other times it may be gross human error. Whatever the cause of error and whoever the responsible agent, there are a few facts that do not merit argument – that error is inherent in any medical process (Gorovitz, 1975) and that errors cause emotional turmoil in both patients and health care providers. These issues deserve equal attention, as do other issues such as how to design safer systems and enhance the quality of care.

Concerns about ethical and emotional issues facing both patients and health care professionals, particularly physicians, are best addressed in light of the current health care environment. Like other issues discussed in this work that would benefit from cultural changes, moral issues need adjustments. The foremost entity that requires evaluation and change is the culture of malpractice suits. This phenomenon in recent times has reached a crisis point (Mello, 2003). It is common to see patients seeking legal redress for their grievances if the eventual outcome of their medical treatment is anything less than perfect. The flawed expectation that health care professionals will be perfect all the time should change. It should be realized that such perfection might not be achievable at all times, since health care delivery involves humans. The unrealistic expectations of patients and increased threats from the current tort system have driven many professionals away from their ethical duty and responsibility to make honest disclosure of mistakes to their patients and their families. This is not a healthy development and will eventually undermine the public trust in the previously noble profession of medicine.

Wu et al. (1991) reported that nearly 76% of house officers have chosen not to disclose a serious error to their patients.

Though research on error disclosure is limited, there is little evidence to suggest that the phenomenon of not disclosing errors to patients has changed over the years. Blendon et al. (2002) reported that only 30% of the patients who had suffered from an error acknowledged having been informed about the error by their physician. Not disclosing an error during the course of patient care may amount to compromising patients' informed consent. Robertson (1987) reported on the necessity of providing information regarding an error in order before a patient can give consent for treatment of injury caused by an error. This lapse on the part of the health care professional breaches an unspoken understanding between the professional and her patients to act in the patient's best interests. Professional ethics aside, Hebert et al. (2001) commented that patients have a right to information about errors by virtue of being respected as a person. Patients have expectations for their physicians regarding disclosure (Gallagher et al., 2003). Patients expect information about what happened, why the error occurred, how the consequences will be managed, and how a recurrence of the error will be prevented.

Health care errors take a severe toll on patients and families. If information about the adverse event is withheld, it only adds to patients' distress and agony. The failure to disclose information on medical mistakes adversely affects patient's decision making, impairs the patient's trust in his physician, and increases the chances for a malpractice suit (Braddock III et al., 1999; Hikson et al., 1992; Witman et al., 1996). In fact, 24% of those who filed malpractice suits did so because they believed that their physician was dishonest and had covered up important information (Hickson et al., 1992). As an incentive to health care professionals, evidence exists that punitive actions may not necessarily be initiated against them if their errors are honestly acknowledged, and appropriately disclosed (Heilig, 1994; Witman et al., 1996).

A noteworthy example is Wighton v. Arnot, an Australian court case from 2005 (New South Wales Supreme Court, 2005). A patient's nerve was severed during an operation, but no investigation or disclosure of this adverse event, which was suspected by the surgeon, ever took place. It was not the treatment that the plaintiff received during the operation – including severing the nerve – however, that was the basis of the negligence assertion. The defendant was alleged to have been negligent in the sense that he had failed to do three things: inform the patient of his suspicion that the nerve had been served; perform an investigation of the suspected incident; and assist the patient in receiving timely corrective surgery. The irony, as Madden and Cockburn (2005) write, is that "the doctor would not have been found liable in negligence had he disclosed the adverse event to the patient."

## 7.5 Patient-physician relations

Another important factor influencing the current health care environment and culture is the poor professional-patient relationship. High workloads and a shortage of medical professionals have drastically reduced the time spent by physicians with their patients in clinical encounters. This has not been accepted well by patients and may be interpreted by them as professional indifference or unavailability (Smith and Forster, 2000). It is suggested that physicians' inadequate communication skills hamper development of relationships with patients, and initiatives to improve these skills are required (Reifsteck,

1998). The current commercialization of health care is also significantly threatening the fiduciary relationship (Crawshaw et al., 1995). These factors, along with the disease factors themselves, lead to an impaired professional-patient relationship and create a deepening sense of mistrust and suspicion. Health care professionals, aware of such feeling among patients, are reluctant to face further awkwardness by disclosing errors and apologizing for them (Smith and Forster, 2000).

Most physicians are not averse to accepting responsibility and apologizing to patients, but fear of expressing regret amounting to legal liability keeps physicians away from this option (Gallagher et al., 2003). So, where exactly does the physician-patient relationship stand in today's environment? The chief medical officer of the United Kingdom, Professor Liam Donaldson, goes to the extent of saying, "We have to see ourselves as the servants of the patients, not their masters" (Donaldson, 2003). A strong statement indeed, but its practicability is open to question. It is better for fostering quality care and patient safety that we stay away from servile debates and consider ourselves as partners with patients in achieving common goals. Improving patient's clinical decision-making abilities may result in better processes and outcomes of care. Improving safety can also be achieved and improved by engaging patients in redesigning processes (Wensing and Elwyn, 2003).

There may be instances when error disclosures are not in the best interest of the patients and may result in psychological harm to the patient. The concept, generally referred to as "therapeutic privilege," should be an exception rather than the rule, exercised only in extreme cases. Health care professionals should routinely disclose errors to patients and their family. Care should be taken that the disclosure occur at the right time, when the patient is in a stable condition and in the right setting (Hebert et al., 2001). Sometimes, challenging situations arise if the error was trivial and may ultimately not cause any harm to the patient. In such instances, Hebert et al. suggested that appropriate disclosure should be made unless evidence exists that the patient may not want to be told about the error.

A study conducted by Tamblyn et al. (2007) indicates that doctors who lack good communication skills are more likely to have complaints filed against them than those who communicate well with their patients. The study analyzed 3,424 physicians licensed in Ontario and Quebec who took the Medical Council of Canada clinical skills examination between 1993 and 1996. The patient-physician communication score in the clinical skills examination was predictive of the number of retained complaints registered.

Given this information, it is unsurprising that Liebman and Hyman (2004) claim that malpractice suits most commonly occur when patients experience both an unexpected adverse event and a lack of empathy from physicians who fail to disclose important information.

## 7.6 The dilemma of an apology

A key recommendation of the various global policies on medical error disclosure is that doctors apologize to the patient, thus soothing anger and lessening suspicion (Cohen, 2000). But doctors and others, though possibly willing to accept responsibility and express regret, may be reluctant to pursue this course if it amounts to admission of guilt or legal liability. Liebman and Hyman (2004) distinguish between two types of apology – "apology of sympathy" and "apology of responsibility." Since some legal jurisdictions consider an apology to be evidence of liability, these authors suggest that doctors and hospital administrators weigh the risks and benefits of an apology beforehand; indeed, it is not uncommon to find that risk managers and hospital attorneys discourage a timely apology for fear of encouraging a lawsuit.

Herein lies a conundrum, in view of the perception that an appropriately worded apology by the doctor can reduce the likelihood of a lawsuit (Mazor et al., 2004). This conflict is partly resolved by measures such as those introduced in Massachusetts and Florida, whereby apologies or expressions of regret to patients are legally protected (Fla Stat §90.4026; Mass Gen Laws ch 233, §23D). Some medical errors are due to system failures (Kalra, 2004), and in these circumstances the doctor may be disinclined to offer an "apology of responsibility." Yet, an insincere apology driven by regulatory standards and institutional policies may carry its own risks.

Although health care organizations are adopting a risk management approach toward medical error that involves taking measures to reduce financial losses due to accidents and injuries and to keep medical malpractice to a minimum, not all organizations agree on how to achieve these ends. One of the main barriers to open and honest disclosure is concern that the financial consequences may be too drastic. Any attempts made by health care organizations to cover up or play down errors in order to mitigate financial losses may, in fact, have the opposite effect. If patients come to see medical institutions as adversaries, then it is more likely that they will seek punitive damages in court, in addition to loss-based ones (Kraman and Hamm, 1999). According to a survey conducted by Hickson et al. (1992), 43% of families who sued their health care provider did so either they suspected that the physician had not disclosed an error or as a way to seek punishment. A separate study discovered that virtually every patient surveyed desired complete disclosure from his or her physician, even for seemingly minor errors (Witman et al., 1996). This second survey also noted that a majority of patients would seek legal recourse if their physicians failed to inform them of moderate to severe medical errors.

A Veterans Affairs Medical Center in Lexington, Kentucky, however, has bucked this trend by adopting a humanistic risk management policy that encourages review of injuries sustained during hospitalization, full disclosure of adverse events, including medical error, and fair compensation for injuries. Compensation primarily takes the form of settlements calculated in terms of actual loss experienced by the patient, including additional medical costs, lost work opportunities, and pain and suffering. Hospital management conduct reviews and, if they find injuries to be the result of medical errors, the patients involved are informed and settlements are proposed. Kraman and Hamm, who studied this hospital (1999), claim that "five settlements involved incidents that caused permanent injury or death but would probably never have resulted in a claim without voluntary disclosure to patients or families."

The Lexington hospital paid out approximately $190,000 each year that Kraman and Hamm studied the hospital. Only one case proceeded to federal trial, and in this case the defendant won. One of the benefits of conducting thorough and timely reviews is that frivolous claims were no longer effective at winning settlements, because it was easier to determine whether or not a potential claimant's case had merit. The hospital could no longer have its hand forced into settling out of court to avoid the cost and work of a malpractice suit, since, after such reviews, it could be confident that it would win such disputes at trial (Kraman and Hamm, 1999).

Although a risk management policy that provides patients with the greatest possible amount of information should also enable patients to maximize liability claims, the Lexington hospital's payouts were moderate when compared to those of similar facilities analyzed by Kraman and Hamm (1999). The authors contribute this primarily to a strengthened fiduciary patient-physician relationship, under which plaintiffs are more willing to negotiate a settlement on the basis of damages incurred by the hospital and choose not to further financially punish their health care providers. Settlements reached out of court are also less costly than the litigation process, which can stretch over many years and involve many expenses, including fees for expert witnesses and litigation teams and incidental expenses.

## 7.7 Barriers to full disclosure

The IOM has identified five barriers to reporting medical errors (Kohn et al., 2000). First, there is a concern for keeping the identities of providers and organizations confidential because of the threat of legal action. Second, medical professionals may believe that the time spent reporting will be wasted because the reported information may not be used; new medical equipment and pharmaceuticals are rigorously tested before they are distributed to medical practitioners, who may believe that possible risks of new technologies are already known, and risky and aggressive interventions are standard practice in health care simply because medical professionals believe that the benefits outweigh the potential risks. Third, there is a lack of education and training in recognizing medical error. Fourth, there is a lack of clear standards, tools, and definitions for reporting these errors. Fifth, there is a lack of compensation for those organizations that undertake the cost of reporting (Kohn et al., 2000).

The risk of legal action, however, may not be the greatest deterrent to open and timely disclosures. Despite the protection conferred on physicians in New Zealand, the number of medical errors reported has not significantly increased (Davis, 2002). As Paterick et al. (2009) note: "One might infer that shame, not blame, inhibits reporting of medical errors in New Zealand. Another reason could be that in the absence of any public or professional knowledge of individuals responsible for medical errors, both the deterrent effect for the future errors and motivation to report are compromised." Anonymity is also important. Phillips et al. (2006) determined that physicians and medical support staff would submit error reports only when personal safeguards were in place.

Moreover, the majority of patients who experience injury due to medical error do not sue. Localio et al. (1991) determined that approximately 1% of patients in their study sample of 30,000 randomly selected New York patients were negligently injured, and only 4% of these injured patients sued. The costs incurred by the medical system in rebutting those who do sue, however, are unreasonably high. A study conducted by Studdert et al. (2006) found that 3% of the claims they examined in which patients sought monetary compensation for an injury allegedly sustained during the medical process were lacking evidence of injury upon an independent review of medical records and claim files. In a further third of claims, while injuries were documented, they were not deemed to be the result of medical errors. The total litigation cost of claims that lack evidence of injury and error, however, account for 13% of total litigation costs. Nevertheless, the study concludes that "portraits of a malpractice system that is stricken with frivolous litigation [claims that lack any serious value] are overblown." The study also claims that "the malpractice system performs reasonably well in its function of separating claims without merit from those with merit and compensating the latter," with only 16% of claims determined to be valid going uncompensated (Studdert et al., 2006). Nevertheless, the overhead cost of litigation remains high. The total cost of litigating the claims was 54% of the compensation paid to plaintiffs (Studdert et al., 2006). The fact that the majority of these costs were incurred in resolving claims in which medical error did transpire requires that the system be made more effective if costs are to be reduced; the average claim was resolved five years after the initial injury occurred. This appears to be an area that needs improvement, rather than seeking tort reforms to block frivolous claims. Although some industry leaders call for a cap on financial damages awarded to patients, such a policy would do nothing to ensure fair compensation to patients. It would also fail to improve their safety, which must remain the fundamental issue.

Former senators Hillary Clinton and Barack Obama (2006) identify four goals that an effective tort system would achieve. It would reduce the rates of preventable adverse events; improve the fiduciary patient-physician relationship by promoting open communication; provide fair compensation for deserving victims of medical error; and reduce liability insurance premiums for health care institutions. If more effective patient safety initiatives can be developed, then fewer malpractice cases will have the opportunity to go to court, reducing the risk of liability and thus driving down insurance premiums. Mandating open and complete disclosure concerning important medical information will also decrease the number of lawsuits filed.

## 7.8 Conclusion

The culture of malpractice suits continues to grow. Suits filed solely for monetary considerations abuse the tort system and set an unacceptable trend (Kalra et al., 2004). Blame and retribution may have their place, but society's interests are best served by creating a trusting environment that promotes honest disclosure of error. To restore trust and perhaps to lower malpractice claims, both the public and health care providers must avoid the "shame and blame" game. Another challenge lies in achieving a balance between a nonpunitive approach to error and the need for a process that includes accountability and suitable compensation for patients. We suggest that a system-based error disclosure program can achieve this balance.

## References

Blendon RJ, DesRoches CM, Brodie M, Benson, et al. Views of practicing physicians and the public on medical errors. N Engl J Med 2002;347:1933–1940.

Braddock III CH, Edwards KA, Hasenberg NM, Laidley TL, Levinson W. Informed decision making in outpatient practice: time to get back to basics. JAMA 1999;282:2313–2320.

Breen, K, N.L. Breast-cancer patients reach $17.5M settlement over botched tests. St. John's Telegram, October 31, 2009. http://www.canada.com/health/breast+cancer+patients+reach+settlement+over+botched+tests/2168107/story.html. Accessed August 6, 2010.

Canadian Broadcasting Corporation. Review finds pathologist's rate of errors within average. June 26, 2008. http://www.cbc.ca/canada/manitoba/story/2008/06/26/pathologist-review.html#socialcomments. Accessed August 6, 2010.

Clinton HR, Obama B. Making patient safety the centerpiece of medical liability reform. N Eng J Med 2005;354:2205–2208.

Cohen JR. Apology and organizations: exploring an example from medical practice. Fordham Urban Law 2000;27:1447–1482.

Crawshaw R, Rogers DE, Pellegrino ED, et al. Patient–physician covenant. JAMA 1995; 273:1553.

Davis P, Lay-Yee R, Briant R, Ali W, Scott A, Schug S. Adverse events in New Zealand public hospitals I: occurrence and impact. N Z Med J 2002;115(1167):U271.

Donaldson, L. The HSJ interview: Professor Sir Liam Donaldson. Rock and role. Interview by Alastair McLellan. Health Serv J. 2003;113:20–21.

Fla Stat §90.4026.

Gallagher TH, Waterman AD, Ebers AG, Fraser VJ, Levinson W. Patients' and physicians' attitudes regarding the disclosure of medical errors. JAMA 2003;289:1001–1007.

Gorovitz S, MacIntyre A. Toward a theory of medical fallibility. Hastings Cent Rep 1975;5:13–23.

Hebert PC, Levin AV, Robertson G. Bioethics for clinicians: 23. Disclosure of medical error. CMAJ 2001;164:509–513.

Heilig S. Honest mistakes. From the physician father of a young patient. Camb Q Healthc Ethics 1994;3:636–638.

Hickson GB, Clayton EW, Gethens PB, Sloan FA. Factors that prompted families to file medical malpractice claims following perinatal injuries. JAMA 1992;267:1359–1363.

Kalra J, Saxena A, Mulla A, Neufeld H, Qureshi M, Massey K. Medical error: a clinical laboratory approach in enhancing quality care [Abstract]. Clin Biochem 2004; 37:732–733.

Kohn LT, Corrigan JM, Donaldson MS, editors. Committee on Quality of Healthcare in America, Institute of Medicine. To Err Is Human: Building a Safer Health System. Washington, DC: National Academy Press; 2000.

Kraman SS, Hamm G. Risk management: extreme honesty may be the best policy. Ann Intern Med 1999;131:963–967.

Liebman CB, Hyman CS. A mediation skills model to manage disclosure of errors and adverse events to patients. Health Affairs 2004;23(4):22–32.

Localio AR, Lawthers AG, Brennan TA, et al. Relation between mapractice claims and adverse events due to negligence: results of the Harvard Medical Practice Study III. NEJM 1991;325:245–251.

Madden B, Cockburn T. Duty to disclose medical error in Australia. Australian Health Law Bulletin 2005;14:13–20.

Mass Gen Laws ch 233, §23D.

Mazor KM, Simon SR, Yood RA, et al. Health plan members' views about disclosure of medical errors. Ann Intern Med 2004;140:409–418.

Mello MM, Studdert DM, Brennan TA. The new medical malpractice crisis. N Engl J Med 2003;348:2281–2284.

New South Wales Supreme Court. 637 (Wighton v Arnot) 2005; BC200504663.

Paterick ZR, Paterick BB, Waterhouse BE, Paterick TE. The challenges to transparency in reporting medical errors. J Patient Saf 2009;5(4):205-209.

Phillips RL, Dobey SM, Graham D, Elder NC, Hickner JM. Learning from different lenses: reports of medical errors in primary care by clinicians, staff, and patients. K Patient Saf 2006;2:140–146.

Reifsteck SW. Difficult physician-patient relationships. Med Group Manage J 1998;45: 46–50, 54.

Robertson GB. Fraudulent concealment and the duty to disclose medical mistakes. Alberta Law Rev 1987;25:215–223.

Smith ML, Forster HP. Morally managing medical mistakes. Camb Q Healthc Ethics 2000;9:38–53.

Studdert DM, Mello MM, Gawande AA, et al. Claims, errors, and compensation payments in medical malpractice litigation. NEJM 2006; 354[19]:2024–2033.

Tamblyn R, Abrahamowicz M, Dauphinee D, et al. Physician scores on a national clinical skills examination as predictors of complaints to medical regulatory authorities. JAMA 2007;298:993–1001.

Wensing M, Elwyn G. Methods for incorporating patients' views in health care. BMJ 2003;326:877–879.

Witman AB, Park DM, Hardin SB. How do patients want physicians to handle mistakes? A survey of internal medicine patients in an academic setting. Arch Intern Med 1996;156:2565–2569.

Wu AW, Folkman S, McPhee SJ, Lo B. Do house officers learn from their mistakes? JAMA 1991;265:2089–2094.

# 8 International laws and guidelines addressing error and disclosure

In any health care process, some error is inevitable. As indicated in the US Institute of Medicine's report *To Err Is Human* (Kohn et al., 2000), the challenge is to cut the rate of error to a minimum. In Canada, various strategies are being applied to this end, and the federal government has established a Patient Safety Institute. The UK likewise has a National Patient Safety Agency. However, in the many countries where efforts are being made to reduce adverse events and errors, a neglected issue is honest disclosure to the patient or family. In this chapter we examine the central issues and dilemmas concerning "apology" and suggest how we might work toward a systematic and effective process.

## 8.1 Introduction

National, intranational, and international initiatives and advances to address disclosure following medical error show a range of attitudes toward patients' right-to-know and toward institutional and health care providers' immunity and responsibility. Differences in legal, social, economic, and health care systems, social and cultural factors, and access to health services, which have historically differed among nations, underpin patient experiences of care and expectations of quality and disclosure.

The rate of adverse events in hospital patients reported by studies worldwide has varied from 3.7% in New York to 11% in UK hospitals and 16.6% in Australian hospitals. In Canada, two recent papers give rates of 5% and 7.5% (Baker et al., 2004; Forster et al., 2004) and the report *Health Care in Canada 2004* states that about 5.2 million Canadians (representing a quarter of the population) have experienced a preventable adverse event either themselves or in a family member (Gagnon, 2004). The wide variation in reported adverse event rates is partly due to differences in study methods and patient selection. Moreover, there is no agreement on what constitutes "preventability." Only a few studies looked at preventability of adverse events as part of their original design (Forster et al. 2004; Vincent et al., 2001; Wilson et al., 1995). But there is now a consensus that, in terms of patient safety, many health systems perform below their potential best.

## 8.2 Disclosing preventable adverse events

What, then, is the argument for being open with patients and families, even when the repercussions may be unpleasant and costly? We have to remember that inappropriate blame attribution that serves only regulatory needs will merely alienate professionals and discourage them from participating in system improvements.

The foremost justification for open acknowledgment of medical errors is to safeguard public trust in the medical profession. The responsibility to disclose medical errors is acknowledged in codes of professional ethics (American Medical Association Council on Ethical and Judicial Affairs, 2000). Physicians have an ethical duty to tell the truth in the fiduciary physician-patient relationship. Communication with patients about error is currently not seen as a common medical practice, a situation that clearly must change. Another argument is that patients have a right to information about errors simply because of the respect due to them as persons, and indeed patients expect doctors to recognize this duty (Gallagher et al., 2003). In addition, failure to disclose an error during the course of patient care may compromise not only patients' autonomy but also their informed consent. For example, disclosure may be essential if a patient is asked to give consent for treatment of injury caused by an error (Robertson, 1987). Thus, failure to disclose information on medical mistakes adversely affects the patient's ability to make intelligent decisions, impairs patient trust in the doctor, and increases the likelihood of a malpractice suit (Hickson et al., 1992).

## 8.3 International progress and initiatives

### 8.3.1 United States

Access to health care underlies patient well-being and safety; President Barack Obama's recent bill to address inequities in health care access through the Affordable Care Act, passed by Congress and signed into law March 2010, will undoubtedly impact patient access to health care (White House Web site, 2010). Many other initiatives have come forth that directly deal with patient safety.

In response to the Institute of Medicine's call for greater transparency and effective patient safety standards, we proposed a "no-fault" model that makes disclosure of adverse events to patients integral to accreditation (Kalra et al., 2004). In 2001, the US Joint Commission on Accreditation of Healthcare Organizations (JCAHO, 2003) announced an "unanticipated outcome" policy that demands that providers or institutions disclose critical events. The only ambiguity concerns the operational definition of an unanticipated outcome, which institutions must decide for themselves. In general, the Joint Commission recommends that doctor conduct the disclosure, though on occasion some other member of the team will be more suitable. Some individual states, among them Pennsylvania, Nevada, and Florida, have in recent times complemented the federal initiatives by imposing a statutory duty on establishments to notify patients in case of an adverse event (Fla Stat §395.1051, 2003; MCare Act. 13 Oa C S §308, 2002; Nev Rev Stat §439.835, 2003).

More recently, the federal Patient Safety and Quality Improvement Act was signed into law on July 29, 2005, as a direct response to the IOM's 1999 report (Patient Safety and Quality Improvement Act, 119 Stat. 424, 2005). The act establishes a confidential reporting structure in which medical professionals can voluntarily report to independent Patient Safety Organizations (PSOs) on events that adversely affect patients. These organizations receive error reports confidentially, having been granted federal legal protection. Only limited information gained by PSOs can be divulged to certain researchers and the Food and Drug Administration. Disclosed information can be used by law enforcement if and only if the discloser has reasonable cause to believe that the disclosed event constitutes a crime. In a court of law, disclosed information can be requested if it contains evidence of a criminal act, is available, and is not reasonably available from any other source. The act specifies that the role of PSOs is to collect, develop, analyze, and maintain patient safety event information. Any person or organization that knowingly violates the confidentiality provisions is subject to penalties.

The act also calls for the creation of a Network of Patient Safety Databases (NPSD), which would provide evidence-based management resources for entities such as health care providers and other PSOs. Using common formats such as definitions and data elements, the NPSD will promote the interchange of information between reporting systems and PSOs.

### 8.3.2 New Zealand

New Zealand has adopted a no-fault compensation policy that entitles patients to payments from the Accident Compensation Corporation if they are injured due to unintended medical errors or any other adverse consequence that results from their medical treatment (Paterick et al., 2009). Medical professionals pay insurance premiums to fund these payments (Davis et al., 2002). In turn, physicians are immune to individual legal liability, as tort claims are prohibited under federal law. The law aims to encourage the disclosure of medical errors between patient and physician, as well as encourage patient compensation claims. While patient claims data are aggregated and publicized, specific information regarding the individuals involved is withheld.

### 8.3.3 Australia

In 2002, a committee of the Australian Council for Safety and Quality in Health Care offered an approach to achieving open and honest communication with patients after an adverse event (Australian Council for Safety and Quality in Health Care, 2002), addressing the interests of consumers, healthcare professionals, managers, and organizations. Like the policy proposed by the US Joint Commission, the draft standard is flexible in allowing development of local policies and procedures. The unique aspect of the Australian draft standard is the integration of disclosure into a risk management analysis and investigation of the critical event. The level of investigation depends on grading of the event according to the extent of injury and the likelihood of its occurrence.

A decision by the New South Wales Supreme Court decreed that a legal duty to disclose medical errors occurs when the knowledge is relevant to possible changes to the patient's care and medical outcome. As Justice J. Bryson remarked: "communication with the patient, both before and after treatment, of the diagnosis, advice about what treatment is proposed, and of a report of what treatment has taken place are all integral and essential parts of treatment. They are essential where the patient is conscious and has the capacity to participate in them, because of the nature of the patient as a person with a right to give or withhold consent to an intervention in his body by another person" (New South Wales Supreme Court, 1994).

The Australian tort system was also reformed to afford medical professionals some protection when they make apologies to patients who have experienced an adverse event. The Civil Liability Act of New South Wales (Civil Liability Act [NSW], 2002), which is similar to those in many other provinces, reads as following:

69 Effect of apology on liability

(1) An apology [an expression of sympathy or regret] made by or on behalf of a person in connection with any matter alleged to have been caused by the person:

    (a) does not constitute an express or implied admission of fault or liability by the person in connection with that matter and

    (b) is not relevant to the determination of fault or liability in connection with that matter

(2) Evidence of an apology made by or on behalf of a person in connection with any matter alleged to have been caused by the person is not admissible in any civil proceeding as evidence of the fault or liability of the person in connection with that matter.

A duty to disclose, however, does not exist in Australia, nor does a legal duty to express sympathy or regret (Madden and Cockburn, 2005).

### 8.3.4 Great Britain

The National Health Services (NHS) in 2003 declared a "duty of candour" whereby doctors and managers must inform a patient of an act of negligence or omission that causes harm (Department of Health, 2003). This is in keeping with the policy of the General Medical Council, a London-based medical regulatory agency, which in its 2001 edition of *Good Medical Practice* exhorts: "If a patient under your care has suffered serious harm, through misadventure or for any other reason, you should act immediately to put matters right, if that is possible. You explain fully to the patient what has happened and the likely long-and-short term effects" (General Medical Council, 2001).

The NHS Redress Act, passed in 2005, was the end product of an examination of how handling and responding to clinical negligence claims made against the NHS could be improved. The previous system, in which negligence claims passed through a costly and complex procedure, provided medical professionals with a deterrent to disclosure, thus preventing improvement in the health care service. The aim of the bill was to "provide a genuine alternative to litigation for less severe cases where there is a qualifying liability in tort arising out of hospital treatment," thereby lowering the burden of litigation of the patient (Department of Health, 2005). Patients no longer suffer the anxiety caused by pursuing litigation. The act also reduces the costs incurred by each side, as neither has to hire competing legal counsel or assume the risk of losing the claim.

### 8.3.5 Canada

In 2002, the Royal College of Physicians and Surgeons of Canada called for health care systems to promote disclosure on safety issues to all partners including patients (National Steering Committee on Patient Safety, 2002), but no uniform Canadian guidelines on the subject are yet in place (see ▶ Fig. 8.1). Reviewing nationwide practices on adverse event disclosure, we found that just a few licensing bodies had ratified policies for disclosure and discussion of negative outcomes during patient care. The College of Physicians and Surgeons of Saskatchewan requires physicians to disclose any adverse events and errors to the patient or his or her representative as soon as possible during care and provides 10 guidelines for purposeful disclosure. The College of Physicians

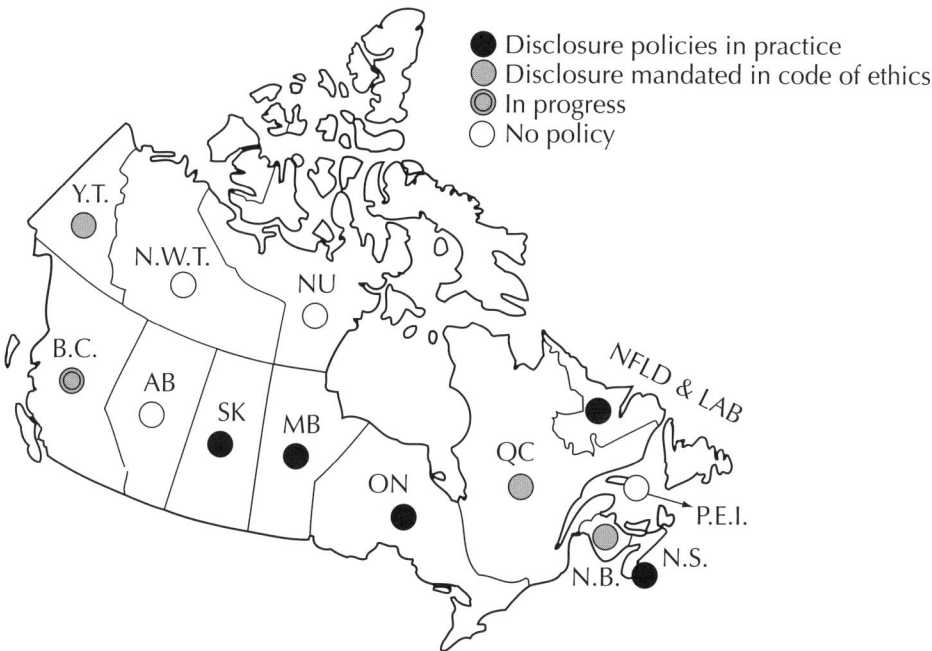

**Fig. 8.1:** Canadian provincial policies on disclosure of harm.

and Surgeons of Manitoba requires physicians to avoid all speculations and to state plain facts as known at the time. In 2003, after lengthy deliberation, the College of Physicians and Surgeons of Ontario approved a policy that made disclosure of harm to patients a standard of practice (Borsellino, 2003), even in circumstances where such disclosure might result in a complaint or a malpractice insurance claim. A special aspect of the Ontario College policy is the guideline for medical trainees (i.e., students or residents), who are advised to report adverse events either to their supervisor or to the "most responsible physician."

The College of Physicians and Surgeons of Quebec has no distinct policy on adverse event disclosure to patients but synthesizes the concept of disclosure in its code of ethics. The College of Physician and Surgeons of Newfoundland and Labrador enacted a disclosure policy in 2006, emphasizing that a disclosure is not an admission of fault or liability. The College in Nova Scotia developed a disclosure policy two years later and also initiated a quality assurance program to complement the policy (see ▶ Tab. 8.1).

Disclosure may also be viewed as part of the informed-consent process. While some Canadian provinces have regulatory initiatives to address disclosure and though many of these are similar in content, they remain isolated because of their nonmandatory nature and an absence of federal and provincial laws. It is suggested that a uniform national policy centered on addressing errors in a nonpunitive manner and respecting the patient's rights to an honest disclosure be implemented (Kalra et al., 2007). In Canada, there is nothing in the nature of the United States Joint Commission initiative that makes

**Tab. 8.1:** Canadian provincial College of Physicians and Surgeons policies on disclosure.

| Province | Key Points |
| --- | --- |
| Saskatchewan | 2002 – Include an apology |
| Manitoba | 2002 – Avoid speculation |
| Ontario | 2003 – Patient can refuse discussion |
| Quebec | Integrated into code of ethics |
| Newfoundland and Labrador | 2006 – Not an admission of fault or liability |
| British Columbia | Prevents an apology as an admission of liability |
| Nova Scotia | 2008 – Implementation of a "just culture" |

disclosure of adverse events a requirement for hospital accreditation. The absence of laws, federal or provincial, mandating adequate disclosure of an adverse event to the patient is a key area of concern.

## 8.4 Conclusion

Clearly, a wide range of responses to patient safety concerns exists in Western nations' health care systems. Canada has much to learn from such progressive initiatives in many countries. Strong examples are present that attempt to balance physician and health care worker responsibility to the patient against unnecessary risk of legal action. Development of international standards may be one way to encourage and promote the continued development of policies in this area. Yet, variation within nations, such as that among the provinces in Canada, epitomizes why regulations may be difficult to enact for this purpose. There is a need for cross-country comparative studies to examine differences between national policies and outcomes to determine best approaches.

## References

American Medical Association Council on Ethical and Judicial Affairs (AMACEJA). Code of Medical Ethics: Current Opinions. Chicago: American Medical Association; 2000.

Australian Council for Safety and Quality in Health Care. Draft Open Disclosure Standard. Standards Australia, XX 1234–2002, Draft v5.2.

Baker GR, Norton PG, Flintoft V, et al. The Canadian Adverse Events Study: the incidence of adverse events among hospital patients in Canada. CMAJ 2004;170:1678–1686.

Borsellino M. Disclosure of harm to be standard of practice. Medical Post 2003;39 (11).

Civil Liability Act 2002 (NSW), §69.

Davis P, Lay-Yee R, Fitzjohn J, Hider P, Briant R, Schug S. Compensation for medical injury in New Zealand: does "no-fault" increase the level of claims making and reduce social and clinical selectivity? J Health Polit Policy Law 2002;27(5):833–854.

Department of Health. Making Amends. London: DoH, 2003.

Department of Health, NHS Redress: Statement of policy. 272080;5787, 2005.

Fla Stat §395.1051 (2003).

Forster AJ, Asmis TR, Clark HD, et al. Ottawa Hospital Patient Safety Study: incidence and timing of adverse events in patients admitted to a Canadian teaching hospital. CMAJ 2004;170:1235–1240.

Gagnon L. Medical errors affect nearly 25% of Canadians. CMAJ 2004;171:123.

Gallagher TH, Waterman AD, Ebers AG, Fraser VJ, Levinson W. Patients' and physicians' attitudes regarding the disclosure of medical errors. JAMA 2003;289:1001–1007.

General Medical Council. Good Medical Practice. 3rd ed. London: GMC; 2001.

Health care: about the new law. http://www.whitehouse.gov/healthreform/healthcare-over view#healthcare-menu7. Accessed October 25, 2010.

Hickson GB, Clayton EW, Githens PB, Sloan FA. Factors that prompted families to file medical malpractice claims following perinatal injuries. JAMA 1992;267:1359–1363.

Joint Commission on Accreditation of Healthcare Organization. Comprehensive Accreditation Manual for Hospitals: The Official Handbook. Chicago, Illinois: JCAHO; 2003.

Kalra J, Massey KL, Mulla A. Disclosure of errors. Health Affairs 2004;23:273–274.

Kalra J, Neufeld H, Mulla A. Disclosure of medical errors: a view through a global lens [Abstract #12]. Clin Invest Med 2007; 30(4S):33.

Kohn LT, Corrigan JM, Donaldson MS, editors. Committee on Quality of Healthcare in America, Institute of Medicine. To Err Is Human: Building a Safer Health System. Washington, DC: National Academy Press; 2000.

Madden B, Cockburn T. Duty to disclose medical error in Australia. Australian Health Law Bulletin 2005;14:13–20.

Mass Gen Laws ch 233, §23D.

Medical Care Availability and Reduction of Error (Mcare) Act. 13 Pa C S §308 (2002).

National Steering Committee on Patient Safety. Building a Safer System – a National Integrated Strategy for Improving Patient Safety in Canadian Health Care. Ottawa, Ontario: NSCOS; 2002.

Nev Rev Stat §439.835 (2003).

New South Wales Supreme Court. Bryson J, October 10 1994; BC9403138.

Paterick ZR, Paterick BB, Waterhouse BE, Paterick TE. The Challenges to Transparency in Reporting Medical Errors. J Patient Saf 2009;5:205–209.

Patient Safety and Quality Improvement Act. 109th Congress. Public Law 109–41. 119 Stat. 424. July 29, 2005.

Robertson GB. Fraudulent concealment and the duty to disclose medical mistakes. Alberta Law Rev 1987;25:215–223.

Vincent C, Neale G, Woloshynowych M. Adverse events in British hospitals: preliminary retrospective record review. BMJ 2001;322:517–519.

Wilson RM, Runciman WB, Gibberd RW, Harrison BT, Newby L, Hamilton JD. The Quality in Australian Health Care Study. Med J Aust 1995;163:458–471.

# 9 The value of autopsy in detecting medical error and improving quality

The autopsy is revered as the gold standard in medical quality assessment, meaning that there exists no better methodology for determining the cause of a patient's death. A physician's original diagnosis of the patient's disease can therefore be determined to be correct if it is concordant with the postmortem diagnosis; if discordant, then the physician made an error. This chapter addresses the ways in which the autopsy contributes to modern medical knowledge and its unique contribution to patient safety.

## 9.1 Introduction

Autopsy examinations have been suggested as one of the methods in quality assessment (Anderson, 1984). Autopsy diagnoses have traditionally been used as the gold standard for determining the cause of death and can provide an invaluable retrospective look into the ability of the medical staff to accurately diagnose the actual cause of death. Despite all of the technical advances in medical and diagnostic modalities, research concerning medical errors in hospitalized populations has consistently revealed high rates of misdiagnoses. The medical literature is replete with studies that suggest discrepancies between antemortem clinical diagnoses and postmortem autopsy diagnoses. Concordance rates of clinical and autopsy diagnoses have been reported to vary from 32% to 77% (see ▶ Tab. 9.1). Utilization of autopsies as part of an error reduction strategy may be a common denominator in improving quality, one that may be used more frequently if we put in place measures to encourage its undertaking.

## 9.2 Error in diagnostic medicine

Diagnostic autopsies can provide an invaluable retrospective look into the ability of medical staff to accurately diagnose the actual cause of death. If the final diagnosis is found to be discordant with the original diagnosis, then a medical error has occurred. We recently (Kalra et al., 2010) conducted a retrospective chart review of 3,416 adults who died as inpatients in a Canadian city over a three-year period. To discern whether the clinical diagnosis and the actual cause of death were the same, Goldman's criteria of misdiagnoses were used (Goldman et al., 1983).

The concordance rate between the clinical and autopsy diagnoses during the studied period was found to be 75.3%. The discordance rate was 20.9%. In a further 3.8% of the cases, our review was deemed to be indecisive, as the cause of the death could not be determined. This high rate of misdiagnoses is not unusual in postmortem studies. A retrospective review of deaths occurring in a Spanish emergency department found

**Tab. 9.1:** Concordance rates of clinical and autopsy diagnoses.

| Study Authors | Year | Study Sample Size | Concordance Rates (%) |
|---|---|---|---|
| Mort and Yeston | 1999 | 149 | 58 |
| Gut et al. | 1999 | 30 | 67 |
| Tse and Lee | 2000 | 332 | 62 |
| Ermenc | 2000 | 1,792 | 50 |
| O'Connor et al. | 2002 | 59 | 51 |
| Coradazzi et al. | 2003 | 156 | 79 |
| Perkins et al. | 2003 | 38 | 45 |
| Combes et al. | 2004 | 167 | 32 |
| Gibson et al. | 2004 | 348 | 52 |

that misdiagnoses or missed diagnoses occurred in 26% of the cases analyzed (Balaguer et al., 1998).

A separate study at another Spanish hospital revealed that more than half of the risk of death could be traced back to clinical care that resulted in adverse events (Garcia-Martin et al., 1997). Although autopsy rates are declining, autopsies remain an important method by which medical error can be detected. Despite the use of modern medical technology to help medical professionals more accurately diagnose patients, including the use MRIs and CT scans, the frequency of discordance has remained largely unchanged for more than 40 years (Lundberg, 1998).

We have suggested a model in which the health care sector either incentivizes the legally responsible party to allow an autopsy or promotes the benefits of autopsies and postmortem findings (Kalra et al., 2010). After all, autopsies contribute to clinical knowledge by extending the scientific and medical understanding of the human body and disease processes. They also aid medical education by allowing medical students and residents to train with greater efficiency by having access to actual human tissues and organs. Lesar et al. (1990) found that there were more prescribing errors – typically a result of inaccurate diagnosis – among first-year postgraduate residents than among other clinicians. Wilson et al. (1998) discovered that the likelihood of error increased in a pediatric intensive care unit when a new doctor joined the rotation. If the quality of medical education can improve, it is likely that the number of inaccurate diagnoses and other medical errors made by new clinicians will in turn decrease.

## 9.3 Missed diagnosis and discordance

Goldman et al. (1983) devised criteria for classifying missed diagnoses. Class I includes missed diagnoses that, if detected prior to death, would have led to a change in management, leading to cure or prolonged survival. Class II refers to missed diagnoses that, if detected, would not have led to a change in management. Classes III and IV refer to missed minor diagnoses involving diseases that were not directly related to the death of the patient. Gibson et al. (2004) added a fifth discordance category in their study of postmortem discrepancies: overdiagnosis, that is, instances in which patients are diagnosed with additional diseases, either principal or secondary, whose treatment would not have altered the outcome in anyway.

It is important to note that misdiagnoses do not always represent errors. They may instead reflect acceptable limits of antemortem diagnoses or atypical clinical presentations. When such atypical presentations occur, however, the autopsy becomes extremely important, as it can reveal the clinical course of rare or unrecorded diseases.

## 9.4 The value of autopsies

Autopsies are valuable in clarifying medico-legal issues, such as whether a death was an accident, a suicide, or a homicide, by uncovering the causes of death. In a review of malpractice suits, when autopsy findings favored the contentions of the plaintiff, the physician was charged with malpractice in 39% of all cases (Bove and Iery, 2002). When autopsy findings favored the defendant physicians' contentions, the physicians were acquitted 100% of the time. Although families of the deceased are often hesitant to allow autopsies to be conducted, a postmortem can assure them that certain actions or inactions taken on their part did not affect the outcome – which is usually the case. It can also detect hereditary and infectious illnesses that other family members may be at risk of developing.

By providing reliable epidemiological data, autopsies also further patient safety in four significant ways. First, they allow pathologists to develop more accurate diagnoses based on recorded signs and symptoms. Perkins et al. (2003) reviewed the autopsies of patients who died in an ICU of a university affiliated hospital and found that the most common diagnostic mistake was a diagnosis of pneumonia; doctors mistakenly diagnosed this illness when in fact the patient had suffered a myocardial infarction. This mistake concerning myocardial infarction was also commonly reported in the study by Gibson et al. (2004) and Combes et al. (2004). Performing additional autopsies would aid research concerning myocardial infarctions and provide valuable information about the signs and symptoms of the disease. Without such information, it is probable that such diagnoses will continue to be missed. Moreover, such information helps to monitor health quality among populations with an increasing proportion of geriatric and obese patients with comorbidities (Fruhbeck, 2004).

The second benefit of autopsies is that they provide data on the clinical course of diseases. Autopsies permit a more invasive examination of the patient than would be allowed if they were alive, allowing for a more accurate charting of the disease process. This is especially true for the brain and other parts of the central nervous system, for which autopsy remains a cornerstone in advancing our scientific understanding (Esiri, 2005). Moreover, new methods of treatment require validation in postmortem stages as well as in living patients. The nature of various diseases and their mechanisms have been established via data extracted from the autopsy, and therapeutic methods have been developed as a result (Grade et al., 2004). For example, autopsy research has recognized a variant of Creutzfeldt-Jakob disease (Chin, 2000; Venters, 2002; Will et al.,

1996). Pediatric technology has also identified the delineation of Reyes syndrome and amoebic meningoencephalitis (Khong and Arbuckle, 2002).

Third, autopsies allow for performance checks, both within the hospital itself and within the health care system overall. If the individual physician in charge of the diagnosis greatly influences misdiagnosis rates at a given hospital, then it's an indication that corrective action must be taken. A study undertaken by Battle et al. (1987), however, indicated that discordance rates are not dependent upon the individual physician. Yet, higher-than-average discrepancy rates in individual hospitals within a health care region are also a cause for concern.

Fourth, autopsies can be used to instruct medical professionals. They can be part of a process to teach anatomy, macroscopic pathology, skills in clinicopathological correlation, the fallibility of medicine, medical bioethics, the process of dying and handling the dead, invasive clinical procedures, medical law, and the importance of health and safety at work (Burton, 2003). The autopsy satisfies these functions because of its visual nature and because of the systems-based, problem-oriented approach adopted by pathologists during the autopsy procedure. Better training produces better doctors, which improves patient safety across the board.

## 9.5 Autopsy decline and strategies to encourage autopsy

Despite the demonstrated value of autopsies as an important quality control tool, it is an unfortunate fact that the number of autopsies conducted in North America is on a decline (Doyle et al., 1990) and despite the Medicare Payment Advisory Committee's recommendation (1999) that autopsies should be more frequently performed, as "information derived from autopsies offers significant potential for use in efforts to reduce errors and improve quality." Several factors affect the rate of performed autopsies. First, unless foul play is suspected, autopsies are not a reimbursable expense; the cost is added to the patient's bill. Families of patients are usually unwilling to take on this expense, and physicians are uneager to push the option. In Belgium, where hospitals assume the costs of all autopsies, some hospitals report autopsy rates of up to 93% (Roosen et al., 2000). Moreover, proper information concerning autopsies is not made readily available to eligible families. One study showed that autopsy literature was available in 4 of 127 surveyed pediatric and medical departments (Rosenbaum et al., 2000).

Religious prohibitions may also be negatively impacting autopsy rates. Hindu and Sikh traditions and teachings require burials without unnecessary delay. The Baha'i, Islamic, Jewish, and Rastafarian faiths, as well as several Christian sects, oppose on religious grounds either the invasive manner of the autopsy or the tissue retention that is often requested. Burton et al. (2004) noted that their study found no relation between the religion, sex, marital status, or educational attainment of the deceased person and the likelihood that consent for the autopsy would be granted, suggesting that religious traditions may be adaptive when circumstances warrant. A national survey of teaching hospitals in the United States revealed that 45% of chief residents had received no training in autopsy consent practice and 82% had received no training in the religious and cultural issues that surround autopsy practice (Rosenbaum et al., 2000).

Nonconventional autopsies exist that may be more agreeable to those with religious or aesthetic concerns (Burton, 2007). Minimally invasive autopsies can be performed in which only blind or targeted percutaneous needle biopsies are taken and only examinations that can be done by endoscope or laparoscope are performed. An autopsy that does not penetrate the body in any way, known as a virtopsy, can be performed via the use of MRI and multislice CT scans. In such a procedure, CT images provide data on the general pathology of the body, which can generate detailed information concerning trauma injuries. MRIs are focused on specific areas of the body, providing data related to tissue, muscles, and organs. Metabolite counts in the brain are measured using MRI spectroscopy to determine time of death.

Despite the general reluctance to perform autopsies, recent reports indicate that up to 46% of relatives may agree to an autopsy if approached sensitively by the physician (Osborn and Thompson, 2001). In a Norwegian hospital, only 9% of deceased patients' next of kin who were approached refused permission for the performance of an autopsy (Ebbesen et al., 2001). Lugli and colleagues (1999), in Switzerland, found that simple measures such as including attending physicians in autopsy discussions with relatives and providing physicians with communication training on how to deal with relatives led to an increase in autopsy rates from 16% in 1997 to 36% in 1998. Another study, however, reported that 80% of families refused permission for autopsies (Combes et al., 2004). It is interesting to note that studies seem to suggest that hospitals in which physicians are more hesitant to seek consent are the ones with the lowest autopsy rates (Burton and Underwood 2003; Burton et al., 2004).

Physicians' hesitancy to seek autopsies may stem from several factors. Doctors may face legal repercussions if clinical diagnostic errors are exposed during the autopsy (Hasson and Gross, 1974). Moreover, many health care professionals believe, despite the high frequency of missed diagnoses, that advances in diagnostic procedures, such as MRIs and CT scans, have reduced the value of autopsies. Chariot et al. (2000) cited perceived long response time, low satisfaction with the content of the autopsy report, and difficulties in obtaining consent from relatives as the three major barriers facing clinicians who request autopsies.

## 9.6 Conclusion

Shojania et al. (2003) suggested that, though the possibility of clinically discordant diagnoses at autopsy has decreased over time, it still remains high enough to warrant the ongoing use of autopsies. The gold standard of diagnosis should not be allowed to disappear. Medical and accreditation institutions that drop minimum autopsy rates from regulatory guidelines are depriving the health care system not only of a valuable quality assessment method but also of information that can help lead to a healthier community. Despite this, in 1986, Medicare removed the minimum autopsy requirement for participating hospitals (O'Leary, 1996); even before that, the Board of Commissioners of the Joint Commission on Accreditation of Hospitals had discontinued its mandate that 20% of hospital deaths undergo autopsies (Reichart and Kelly, 1985). These systemwide changes have reduced the likelihood that autopsies will be conducted.

## References

Anderson RE. The autopsy as an instrument of quality assessment. Classification of premortem and postmortem diagnostic discrepancies. Arch Pathol Lab Med 1984;108:490–493.

Balaguer MJV, Gabriel BF, Braso AJV, Nunez SC, Catala BT, and Labios GM. El papel de la autopsia clinica en el control de calidad de los diagnosticos clinicos en una unidad de urgencias. Anales de Medicina Interna 1998;15:179–182. [in Spanish]

Battle RM, Pathak D, Humble CG, et al. Factors influencing discrepancies between premortem and postmortem diagnoses. JAMA 1987;258:339–344.

Bove KE, Iery C. The role of autopsy in medical malpractice cases. Arch Pathol Lab Med 2002;126:1023–1031.

Burton JL. The autopsy in modern undergraduate medical education: a qualitative study of uses and curriculum considerations. Med Educ 2003;37:1073–1081.

Burton JL, Clinical, educational, and epidemiological value of autopsy. Lancet 2007;369: 1471–1480.

Burton JL, Underwood JC. Necropsy practice after the "organ retention scandal": requests, performance, and tissue retention. J Clin Pathol 2003;56:537–541.

Burton EC, Phillips RS, Covinsky KE, et al. The relation of autopsy rate to physician's beliefs and recommendations regarding autopsy. Am J Med 2004;117:255–261.

Chariot P, Witt K, Patot V, et al. Declining autopsy rate in a French hospital. Arch Pathol Lab Med 2000;124:739–735.

Chin JE, editor. Creutzfeld-Jakob disease. In: Control of Communicable Diseases Manual. Washington, DC: American Public Health Association; 2000:183–186.

Combes A, Mokhtari M, Couvelard A, et al. Clinical and autopsy diagnoses in the intensive care unit: a prospective study. Arch Intern Med 2004;164:389–392.

Coradazzi AL, Morganti AL, Montenegro MR. Discrepancies between clinical diagnoses and autopsy findings. Braz J Med Biol Res 2003;36:385–391.

Doyle YG, Harrison M, O'Malley F. A study of selected death certificates from three Dublin teaching hospitals. J Public Health Med 1990;12:118–123.

Ebbesen J, Buajordet I, Erikssen J, et al. Drug-related deaths in a department of internal medicine. Arch Intern Med 2001;161:2317–2323.

Ermenc B. Comparison of the clinical and postmortem diagnoses of the causes of death. Forensic Sci Int 2000;114:117–119.

Esiri MM. Why do research on human brains? In: Gunning J, Holm S, eds. Ethics Law and Society. Aldershot: Ashgate; 2005:33–39.

Fruhbeck G. Death of the teaching autopsy: advances in technology have not reduced the value of the autopsy. BMJ 2004;328:165–166.

Garcia-Martin M, Lardelli-Claret P, Bueno-Cavanillas A, Luna-del-Castillo JD, Espigares-García M, Gálvez-Vargas R. Proportion of hospital deaths associated with adverse events. Jour of Clin Epidem 1997;50:1319–1326.

Gibson TN, Shirley SE, Escoffery CT, Reid M. Disrecpancies between clinical and postmortem diagnoses in Jamaica: a study from University Hospital of the West Indies. J Clin Pathol 2004;57:980–985.

Goldman L, Sayson R, Robbins S, Cohn LH, Bettmann M, Weisberg M. The value of the autopsy in three medical eras. N Engl J Med 1983;308:1000–1005.

Grade MH, Zucoloto S, Kajiwara J, Fernandes M, Couto LG, Garcia SB. Trends of accuracy of clinical diagnoses of the basic cause of death in a university hospital. J Clin Pathol 2004;57:369–373.

Gut AL, Ferreira AL, Montenegro MR. Autopsy: quality assurance in the ICU. Intensive Care Med 1999;25:360–363.

Hasson J, Gross H. The autopsy and quality assessment of medical care. Am J Med. 1974;56:137–140.

Kalra J, Entwistle L, Suryavanshi S, Chadha R. Quality assessment using concordance and discordance rates in medical findings. Clin Govern 2010;15:128–133.

Khong TY, Arbuckle SM. Perinatal pathology in Australia after Alder Hey. J Paediatr Child Health, 2002; 38:409–411.

Lesar TS, Briceland LL, Delcoure K, Parmalee JC, Masta-Gornic V, Pohl H. Medication prescribing errors in a teaching hospital. JAMA 1990; 263:2329–2334.

Lugli A. Anabitarte M, beer JH, Effect of simple interventions on necropsy rate when active informed consent is required. Lancet 1999;354:1391.

Lundberg GD. Low-tech autopsies in an era of high-tech medicine: continued value for quality assurance and patient safety. JAMA 1998;280:1273–1274.

Medicare Payment Advisory Committee. Report to Congress: Selected Medicare Issues. Chapter 3: addressing health care errors under Medicare. Washington, DC:1999:47–50.

Mort TC, Yeston NS. The relationship of premortem diagnoses and postmortem findings in a surgical intensive care unit. Crit Care Med 1999;27:299–303.

O'Connor AE, Parry JT, Richardson DB, Jain S, Herdson PB. A comparison of the antemortem clinical diagnosis and autopsy findings for patients who die in the emergency department. Acad Emerg Med 2002;9:957–959.

O'Leary DS. Relating autopsy requirements to the contemporary accreditation process. Arch Pathol Lab Med 1996;129:763–766.

Osborn M, Thompson EM. What have we learnt from the Alder Hey affair? Asking for consent would halt decline in voluntary necropsies. BMJ 2001;322:1542–1543.

Perkins GD, McAuley DF, Davies S, Gao F. Discrepancies between clinical and postmortem diagnoses in critically ill patients: an observational study. Crit Care 2003;7:129–132.

Reichart CM, Kelly VI. Prognosis for the autopsy. Health Aff 1985;4:82–92.

Roosen J, Franse E, Wilmer A. Comparison of premortem clinical diagnoses in critically ill patients and subsequent autopsy findings. Mayo Clin Proc 2000;75:262–267.

Rosenbaum GE, Burns J, Johnson J, Mitchell C, Robinson M, Truog RD. Autopsy consent practice at US teaching hospitals: result of a national survey. Arch Intern Med 2000;160: 374–380.

Shojania KG, Burton EC, McDonald KM, Goldman L. Changes in rates of autopsy-detected diagnostic errors over time: a systematic review. JAMA 2003;289:2849–2456.

Tse GM, Lee JC. A 12-month review of autopsies performed at a university-affiliated teaching hospital in Hong Kong. Hong Kong Med J 2000;6:190–194.

Venters GA. New variant of Creutzfeldt-Jakob disease – the epidemic that never was. BMJ 2002;323:858–861.

Will RG, Ironside JW, Zeidler M, et al. A new variant of Creutzfeldt-Jakob disease in the UK. Lancet 1996;347:921–925.

Wilson DG, McArtney RG, Newcombe RG, et al. Medication errors in pediatric practice: insights from a continuous quality improvement approach. Eur J Ped 1998;157:769–774.

# 10 Total quality management, six-sigma, and health care

The attaining of industrial-grade quality performance by the health care sector is still far from achieved. The industrial sectors have been far ahead of health care in enhancing quality. This wide chasm can be attributed to many factors, particularly the complexity of medicine and disease processes. In addition, today's health care delivery is largely dependent on a complex set of internal systems working smoothly and efficiently in a coherent manner (Lanham et al., 2003). Unlike most health care areas, the industrial sectors have fairly well established protocols and processes in which every step is precisely defined and well controlled, with little or no variability. Another major difference between these sectors is the slow response shown by health care administrators in accepting and adapting to new quality improvement initiatives. The health care delivery system is constantly changing; it is fraught with newer risks every day, and in such a setting the traditional quality improvement techniques such as Continuous Quality Improvement (CQI) and Total Quality Management (TQM) are inadequate and the need for embracing newer quality management models is inevitable and essential. It is for this reason that we propose the six-sigma program. In this chapter, we illustrate its usage, primarily in the clinical laboratory environment.

## 10.1 Introduction

The quality of health care has been a primary concern of many governments worldwide. The IOM conducted a National Roundtable on Healthcare Quality as early as 1996 to deliberate quality-of-health and health care issues in the United States. The roundtable comprised 20 representatives of the private and public sectors, medical and nursing practitioners, academicians, business professionals, patient advocates, media persons, and health administrators. The role of the National Health Service (NHS) in the United Kingdom in improving the quality of patient care was outlined clearly in the government white paper *The NHS: Modern and Dependent* (Crook, 2002). In this paper, the concept of clinical governance has been vividly set out. Clinical governance is defined as a "system through which [NHS] organizations are accountable for continuously improving the quality of their services and safeguarding high standards of care by creating an environment in which excellence in clinical care will flourish" (Scally et al., 1998). Chassin and Galvin (1998), making a statement on the IOM National Round Table, wrote: "Problems in healthcare quality are serious and extensive; they occur in all delivery systems and financing mechanisms." The authors further noted that "Americans bear a great burden of harm because of these problems, a burden that is measured in lost lives, reduced functioning and wasted resources," and called for urgent action.

## 10.2 New issues, newer solutions

The present need is for health care services to build on the past successful strategies and adapt to the modern challenges of managed health care, competition, and increasingly complex health care delivery systems. This task is best accomplished by increasing the focus on improving the processes themselves. In this scenario, the industrially tested and proven six-sigma methodology, through its statistical component and process-focused approach, provides an optimal solution to ease at least some health care quality woes. In some ways, the slow and guarded response of health care sectors in implementing the six-sigma strategy may be a blessing in itself. The industrial and manufacturing sectors have tried, tested, and repaired any kinks in the methodology. As a result, it has gone through the grind and been refined; with a proven record, the six-sigma approach is now ready for application in enhancing the quality and patient safety in health care.

The six-sigma methodology is an industrial quality improvement tool. The industrial sector, particularly Motorola – six-sigma is a federally registered trade mark of Motorola, General Electric (GE) and Allied Signal – has employed six-sigma strategies with remarkable gains in efficiency, client and customer satisfaction, and overall profitability (Harry, 1998). Unlike other quality initiatives borrowed by the health care sector from the industrial sectors (e.g., TQM and CQI), six-sigma improvement provides sustained strategic achievements and long-lasting benefits. The six-sigma philosophy is based on a reduction of variation in a process, as well as customer-oriented and data-driven decisions.

Sigma ($\sigma$), a Greek letter, is used to describe variability in a process. In the six-sigma methodology, the unit used is defects per unit. A sigma value indicates the frequency of defects occurring in a process. Therefore, a higher sigma value translates to fewer defects, and a lower sigma value means a greater number of defects. A process is said to be performing at "world-class" levels when it is functioning at levels of six-sigma (Harry, 1998). In other words, a process performing at six-sigma level translates into a phenomenal 3.4 defects per million (DPM) opportunities, the practical limit to perfection. The present day health care services are functioning at only three-sigma and in some cases four-sigma levels that translate roughly into 66,807 and 6,210 DPM opportunities, respectively (see ▶Tab. 10.1). The only health care sector that has been close to achieving six-sigma performance is anesthesia, with mortality rates – taken as defects – as low as five per million opportunities (Eichhorn, 1989). Though six-sigma quality performance may not be achievable by all, setting a goal of six-sigma performance surely is. A six-sigma performance aims at an overall improvement in the performance of the process, and if this is set as a fundamental goal in health care services, we start getting closer to six-sigma level, thereby improving the performance of the process exponentially. It has been suggested that reaching a rate of 3.4 DPM opportunities is less important than developing a process for evaluating error rates and bringing about systematic changes that increase reliability (Johnstone et al., 2003).

Improving health care quality to six-sigma levels becomes imperative when one considers the percentage of the population that uses health care services. With such a large denominator and millions of health care events occurring every day, even a minuscule percentage of errors represent a large number. On the same front, it is worth considering the fact that even a small error may result in catastrophic consequences to a patient's health. The current health care system is content if its process functioning lies within

**Tab. 10.1:** Levels of sigma performance and corresponding defects per million.

| Sigma Level | DPM Opportunities |
|---|---|
| 6 | 3.4 |
| 5 | 233 |
| 4 | 6,210 |
| 3 | 66,807 |
| 2 | 308,537 |
| 1 | 690,000 |

±2 standard deviations (*SD*) of the mean. In a Gaussian distribution, this would result in only a 4.5% defect rate, but considering the potential for health care usage, this would translate into an appalling 45,400 DPM opportunities. These figures would be of little solace to an already ill patient. Clinical diagnostic laboratories are content if their results enclose ±2*SD* or ±3*SD* limits. In other words, they find defect rates of 45,400 DPM opportunities and 2,700 DPM opportunities (see ▶Tab. 10.2) to be acceptable performance (Harry et al., 2000).

Many claim that little is gained from improving a process performance beyond the five-sigma, or 233 DPM, level. It is felt that six-sigma method applications can actually tolerate small shifts in the process mean without increasing the defect rate that significantly. With a six-sigma process, we are assured that the process is still producing results within the desired specifications and with low defect rates. The six-sigma process provides an added advantage in that it is easily monitored with any QC procedure, unlike a process at five-sigma or lower sigma levels, where the choice of QC procedure is more important.

In any process, variation is inherent. It is variation in the process that creates the opportunities for errors to happen, and therefore it should be seen as the "enemy" (Lanham, 2003). Walter Shewhart (1980) described two types of variation: common cause and special cause. Common-cause variations are intrinsic to a process and require action on the process itself to decrease the variation, whereas special-cause variation occurs due to factors extrinsic to the process, which require identification and action. The key lies in minimizing these variations and producing a stable process. These stable processes exhibit common-cause variations, which are best reduced by correcting the underlying process (Mohammed et al, 2001). It is the variation in a process that has to be minimized and controlled to achieve high-quality results. Reduction in variation is also a core concern in clinical governance. Shewhart (1980) also devised control charts, a graphical methodology for differentiating between the two types of variation. The defects occurring through common-cause variation fall within the upper and lower lines of the graph (control limits), and special-cause variation are represented by the data points that fall outside the control limits. He suggested using limits of three-sigma from the mean. If beyond these points, it was suggested that the process required correction. If one were to apply the three-sigma limit for accepting a process, it would translate into 66,807 DPM opportunities.

**Tab. 10.2:** Gaussian distribution in terms of defects per million opportunities.

| Gaussian Distribution | DPM Opportunities |
|---|---|
| >2SD | 45,400 |
| >3SD | 2,700 |
| >4SD | 63 |
| >5SD | 0.6 |
| >6SD | 0.002 |
| >7SD | $3 \times 10^{-6}$ |

## 10.3 The six-sigma structure

The central theme of any six-sigma project is improving a process. It is suggested that any organization use six-sigma strategies on a smaller scale before implementing six-sigma in a big way. This provides an opportunity to learn the methodology and allows better implementation of bigger and more important projects ahead.

Every organization is different and has its own unique demands and structure. This justifies the different six-sigma approaches employed by organizations. Whatever the organizational approach, the foundation for deploying a successful six-sigma strategy is a combination of sound and effective infrastructure and strong management support. The infrastructure itself may determine the scope and impact of six-sigma strategies in enhancing quality and profits.

Another important component is well-trained personnel, who after a short period of training assume various roles as Champions, Master Black Belts, Black Belts, and Green Belts. The personnel with these unfamiliar and intimidating titles play key roles in the actual implementation of six-sigma projects. The intimidating titles may simply be a reflection of the powerful influence of the six-sigma approach on the quality of a process.

Champions are at the higher end of the hierarchy and are usually high-level executives or division heads who are fully responsible for quality issues and quality improvement of the organization. Their role is particularly critical in the initial stages of six-sigma implementation, where the project may experience technical and administrative hitches. Master Black Belts, as the name suggests, are involved in training the Black Belts. Besides training, they provide technical consultation and leadership to Black Belts. Master Black Belts are experts in six-sigma analytical tools and have a critical role in sustaining the momentum of change, quality enhancement, and cost savings. The Black Belts are also involved full time with six-sigma projects and direct these projects, focusing on finding defects and eliminating them totally from the process. It can be safely argued that a Black Belt's role is the most critical of all, because they are the inspiration and driving force behind all the process improvements. The Black Belts provide leadership to many six-sigma projects in a year. Green Belts also receive six-sigma training but are involved with the projects only part time as an additional duty, while they continue in their regular jobs and accompanying normal duties. The Green Belts do, however, play a crucial role in bringing the concept and analytical tools of six-sigma training directly to everyday activities of the workplace and the process. It is principally for this reason that organizations desiring greater success in the six-sigma projects think about training a large segment of their workforce to be Green Belts. Six-sigma implementation involves every level of an organizational hierarchy, with the top level providing the leadership and the bottom level driving the whole process.

The performance improvement methodology or model used in six-sigma is most often "DMAIC" (Define, Measure, Analyze, Improve, Control). Each letter stands for one of the different stages involved in the implementation of a six-sigma strategy. Though other methodologies for implementing six-sigma strategies also exist, they are practiced by only a few companies and hence are discussed briefly later. When six-sigma is referred to, it is invariably the DMAIC methodology that is being mentioned (see ▶ Fig. 10.1). This is a stepwise graded approach for enhancing the quality of a process to produce the desired goals. The various stages are:

-  Define – the problem in terms of the process, goals the project intends to achieve, custom deliverables, and any other components essential to quality.
-  Measure – the process quantitatively, which is best achieved by data gathering, which helps in assessing the current performance levels and comparison with the best practices.
-  Analyze – the process by using a root-cause analysis approach to determine where the problem is originating and which problem is most responsible for the deteriorating quality of the process.
-  Improve – the process by eliminating the defects through identification of causes.
-  Control – the process so that the improvements are sustained and defects do not re-emerge later.

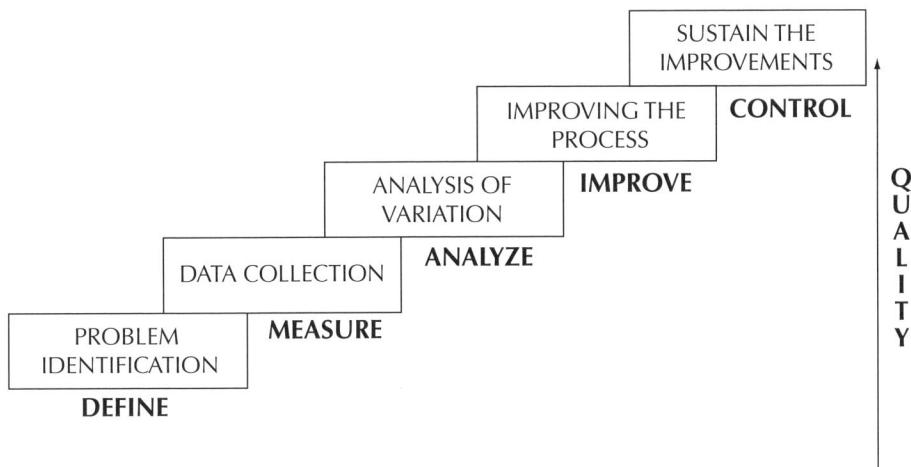

**Fig. 10.1:** The stepwise "DMAIC" approach to enhancing quality.

It should be re-emphasized here that using the DMAIC approach to the six-sigma quality improvement strategies aids in achieving a very low number of DPM opportunities, ideally equal to 3.4 DPM opportunities, which betters 99% performance levels by being 99.9997% perfect all the time. Such a high level of perfection eventually means higher efficiency, leading to reduction in costs, efforts, and time expended and to improved overall client satisfaction.

Apart from the DMAIC methodology, which is essentially used for a preexisting process that is defective, the DMADV (Define, Measure, Analyze, Design, Verify) methodology is used when a new process is being developed or when the DMAIC process has failed to correct a defect. In the DMADV approach, the goals and methods are similar to those in the DMAIC approach – that is, to reduce defects to levels below 3.4 DPM opportunities and data driven, respectively. The difference lies only in the last two steps. In DMADV, they are Design (create a detailed design of the process to meet customer needs) and Verify (check the design performance and its ability in meeting customer needs).

There are other less commonly used methodologies, such as DMADOV (Define, Measure, Analyze, Design, Optimize, Verify), a variant of DMADV; DCCDI (Define, Customer Concept, Design, Implement); IDOV (Identify, Design, Optimize, Validate); and DMEDI (Define, Measure, Explore, Develop, Implement).

## 10.4 Six-sigma in clinical diagnostic laboratories

Like any other system, clinical diagnostic laboratories face numerous barriers in developing and implementing a quality agenda. However, the precise knowledge and delineation of each and every step in a laboratory process and prior experience with using and analyzing statistical data for quality improvement activities allow for some comfort. While clinical laboratories have become a larger, more complex, and more automated area of health care, one in which clinicians must become, to a large extent, system engineers, they also have the advantage of relatively well defined laboratory processes and prior experience with using and analyzing statistical data for quality improvement activities. This advantage, which is not enjoyed by any other health care sector, offsets some of the disadvantages faced by clinical diagnostic laboratories.

A principal barrier to quality is the inadequacy of research in providing a universally accepted definition of error. This makes it difficult to comprehend what actually constitutes a "tolerable error" in clinical laboratories. Clinical diagnostic laboratories work in an atmosphere of cost constraints, which pose a second barrier. This, in turn, does not bode well for quality promotion and enhancement activities. A third barrier is the attitude of laboratory personnel toward quality. An already overworked and tired laboratory workforce views the quality aspects of a laboratory process as "extra work" that interferes with their actual jobs. Embedding quality aspects in the work culture and offering incentives for quality promotion activities and achievements may help bring about a transformation in attitude. A fourth barrier may be a sense of complacency toward quality. Adopting and developing newer quality improvement tools and methodologies may help us realize how far we are from achieving ideal rates of quality.

The most important barrier, however, is the limited control exercised by clinical diagnostic laboratories in the most influential part of the laboratory testing process,

**Fig. 10.2:** Factors driving quality in clinical diagnostic laboratories.

the preanalytical phase. As reviewed earlier, a majority of errors occur in this stage. Because a laboratory analyzes what is delivered to it, preanalytical factors have a direct bearing on the analytical and postanalytical stages of the testing process. This problem is not restricted to the centralized laboratory model but applies also to point-of-care testing (POCT). In a well-reviewed paper on the various aspects of POCT, St-Louis has stressed the demands that the POCT presents in terms of quality and the importance of QA to address all phases of a test performance (St-Louis, 2000).

Six-sigma can be applied widely in all three stages of a clinical diagnostic laboratory testing process. In the preanalytical stage, it can be used to enhance quality of information on requisitions, patient identification, and specimen collection and transportation. In the analytical stage, it can find applications in reducing laboratory testing errors and avoiding misinterpretation, misreading, and misjudging of the results. In the postanalytical stage, it can be used successfully to reduce the turnaround time for obtaining the results.

Previous research in this area has failed to provide us with clear directions for improving quality in our laboratories. It has, rather inadvertently, focused on descriptive statistics and fallen short of exposing the real underlying issues of quality failures. These failures and barriers offer abundant opportunities for further research and development of processes that are efficient and of high quality. In summary, quality in clinical laboratories is driven by application of data-driven approaches and evidence-based practices (see ▶Fig. 10.2). This approach helps in setting up professional, high-quality standards and, coupled with education and training, helps transform a laboratory culture into a quality-conscious setting.

## 10.5 Conclusion

It is imperative for the health care sector in general and for clinical diagnostic laboratories in particular to promote and develop a culture of safety with the aid of modern quality management techniques and tools. Today's quality assurance and improvement activities in clinical laboratories are governed by the Clinical Laboratory Improvement Amendments of 1988 (CLIA '88) and Joint Commission on Accreditation of Healthcare Organizations (JCAHO) guidelines. However, it must be mentioned that the influence of CLIA '88 and JCAHO guidelines is largely confined to clinical laboratories in the United States and may not apply to clinical laboratories in Canada and elsewhere. The criteria set out by CLIA '88 and JCAHO, though highly effective, are not very demanding for analytical performance and are based on two-sigma to three-sigma process goals only.

The goals of six-sigma quality are impressive and set demanding standards that appear to be more compatible with patient safety. In addition, the present philosophy of quality assurance being defined as "find a problem, fix a problem" is not feasible; significant improvements in laboratory performance call for more systematic approaches. The six-sigma concept provides an opportunity for major improvements and helps achieve the vision of ultimate quality to deliver error-free and timely clinical diagnostic laboratory services.

## References

Chassin MR, Galvin RW. The urgent need to improve health care quality. Institute of Medicine National Roundtable on Health Care Quality. JAMA 1998;280:1000–1005.

Crook M. Clinical governance and pathology. J Clin Pathol 2002;55:177–179.

Eichhorn JH. Prevention of intraoperative anesthesia accidents and related severe injury through safety monitoring. Anesthesiology 1989;70:572–527.

Harry MJ. "Six sigma: a breakthrough strategy for profitability." Quality Progress 1998;31:60–64.

Harry M, Schroeder R. The Breakthrough Management Strategy Revolutionizing the World's Top Corporations. New York: Doubleday; 2000.

Johnstone PA, Hendrickson JA, Dernbach AJ, et al. Ancillary services in the health care industry: is Six Sigma reasonable? Qual Manag Health Care 2003;12:53–63.

Lanham B. Beth Lanham on Six Sigma in healthcare. Interview by Luc R. Pelletier. J Healthc Qual 2003;25:26–27, 37.

Mohammed MA, Cheng KK, Rouse A, Marshall T. Bristol, Shipman, and clinical governance: Shewhart's forgotten lessons. Lancet 2001;357:463–467.

Scally G, Donaldson LJ. The NHS's 50 anniversary. Clinical governance and the drive for quality improvement in the new NHS in England. BMJ 1998;317:61–65.

Shewhart WA. Economic control of quality of manufactured product. New York: D Van Nostrand; 1931. (Reprinted by ASQC Quality Press, 1980).

St-Louis P. Status of point-of-care testing: promise, realities, and possibilities. Clin Biochem 2000;33:427–440.

# Index

References to tables are indicated with an italic *t*.